ADVANCES IN MULTI-PHOTON PROCESSES AND SPECTROSCOPY

ADVANCES IN MULTI-PHOTON PROCESSES AND SPECTROSCOPY

Edited by S H Lin

Volume 1

Contents: Atomic Multiphoton Processes by *J. H. Eberly and J. Krasinski;* Some Studies on Laser Multiphoton Ionization and Multiphoton Ionization Dissociation of Polyatomic Molecules by *D. A. Gobeli, J. J. Yang and M. A. El-Sayed;* Laser-Induced Molecular Dynamics: Rate Processes in the Gas Phase and at Solid Surfaces by *J. T. Lin, M. Hutchinson and T. F. George;* Multiphoton Processes by Visible and UV Lasers by *I. Tanaka and M. Kawasaki;* Applications of Raman Spectroscopy to Structural and Conformational Problems by *J. Laane;* Theory of Laser-Stimulated Surface Processes: Master Equation Approach by *B. Fain, A. R. Ziv, G. S. Wu and S. H. Lin.*

Volume 2

Contents: Theory of Molecular Multiphoton Transitions by *Y. Fujimura;* Photochemistry, Photophysics and Spectroscopy of Molecular Infrared Multiple Photon Excitation by *J. S. Francisco and J. I. Steinfeld;* Dynamics and Symmetries in Intense Field Multiphoton Processes: Floquet Theoretical Approaches by *Shih-I Chu;* Time-Resolved Resonance Raman Spectroscopy by *W. Hub, S. Schneider and F. Dorr;* Detection and Spectroscopy of Methyl and Substituted Methyl Radicals by Resonance Enhanced Multiphoton Ionization by *M. C. Lin and W. A. Sanders.*

Volume 3

Contents: Multiphoton Processes in Frequency-Modulated System by *E. Hanamura*; Photodesorption by Resonant Infrared Laser-Adsorbate Coupling — A Review of the Theoretical Approaches by *P. Piercy, Z. W. Gortel and H. J. Kreuzer.*

ADVANCES IN MULTI-PHOTON PROCESSES AND SPECTROSCOPY

Volume 4

Edited by

S H Lin

Department of Chemistry
Arizona State University
Tempe, Arizona 85287, USA

World Scientific

Singapore • New Jersey • Hong Kong

Published by

World Scientific Publishing Co. Pte. Ltd.
P.O. Box 128, Farrer Road, Singapore 9128

U. S. A. office: World Scientific Publishing Co., Inc.
687 Hartwell Street, Teaneck NJ 07666, USA

Library of Congress Cataloging-in-Publication Data

Advances in multi-photon processes and spectroscopy — Vol. 1
  — Singapore : World Scientific, c1984-
    v. : ill. : 23 cm.
  Includes bibliographies.
  Editor: 1984-   S. H. Lin.

  1. Multiphoton processes–Collected works. 2. Spectrum analysis--
Collected works. 3. Laser spectroscopy–Collected works. 4. Molecular
spectra–Collected works. I. Lin, S. H. (Sheng Hsien), 1937-
II. Title: Multiphoton processes and spectroscopy.
QD96.M65A38   543'.0858–dc19   86-643116
                                            AACR 2   MARC-S

Library of Congress   [8610]

ADVANCES IN MULTI-PHOTON PROCESSES AND SPECTROSCOPY VOL. 4

Copyright © 1988 by World Scientific Publishing Co Pte Ltd.

ISSN 0218-0227
ISBN 9971-50-577-0

Printed in Singapore by Utopia Press.

PREFACE

In view of the rapid growth in both experimental and theoretical studies of multiphoton processes and multiphoton spectroscopy of atoms, ions and molecules in chemistry, physics, biology, materials sciences, etc., it is desirable to publish an Advanced Series that contains review papers readable not only by active researchers in these areas but also by those who are not experts in the field but intend to enter the field. The present series attempts to serve this purpose. Each review article is written in a self-contained manner by the experts in the area so that the readers can grasp the knowledge in the area without too much preparation.

The topics covered in this volume are "Progress in Resonance Enhanced Multiphoton Ionization Spectroscopy of Transient Free Radicals", "Time-Resolved Resonance Raman Spectroscopy", "Resonantly Enhanced Multiphoton Ionization — Photoelectron Spectroscopy as a Probe of Molecular Photophysics and Photochemistry" and "Two-Color Double Resonance Spectroscopy for the Study of High Excited States of Molecules". The editor wishes to thank the authors for their important contributions. It is hoped that the collection of topics in this volume will prove useful, valuable and stimulating not only to active researchers but also to other scientists in the areas of biology, chemistry, materials science and physics.

S. H. Lin

CONTENTS

TWO-COLOR DOUBLE RESONANCE SPECTROSCOPY
FOR THE STUDY OF HIGH EXCITED STATES OF MOLECULES

Mitsuo Ito and Masaaki Fujii

Department of Chemistry, Faculty of Science,
Tohoku University, Sendai 980, Japan

1. Introduction

The high excited states of molecules attract a great deal of interest from their importance in physics and chemistry in high energy region[1]. Traditionally, vacuum ultraviolet absorption spectroscopy has been used for the study of high excited state molecules. Recently, the study by this spectroscopy is making rapid progress by introduction of synchrotron radiation as a new light source for vacuum ultraviolet energy region. The traditional method is based on the optical transition of a ground-state molecule by the absorption of one-photon; therefore, it requires a high energy photon to reach the high excited state. Instead of the use of the high energy photon, more than two photons of low energies may be used by the utilization of multi-photon absorption. Although the cross section of multiphoton absorption is much smaller than that of one-photon absorption, powerful lasers made the multiphoton absorption spectroscopy a practical means for the study of the high excited states of a molecule. In particular, the multiphoton ionization spectroscopy first developed by Johnson was proved to be a powerful means for detecting the high excited states of a molecule with laser light of low energy visible or ultraviolet photon[2]. In this spectro-scopy, a molecule is excited to a high excited state by the absorption of two or more photons from the ground-state molecule. The excited state molecule absorbs another one or more photon(s) to reach the ionization continuum; there, the molecular ion is generated. When the ions are detected while scanning the laser frequency, one obtains a two or more photon resonant multiphoton ionization spectrum, which practically coincides with the two-(or more) photon absorption spectrum from the ground state to the high excited state. Since ions can be detected with a high sensitivity, the multiphoton ionization spectroscopy is a highly sensitive spectroscopic means. However, in this spectroscopy, it often happens that the spectra with different resonance states appear in the same spectral region measured by one-photon energy; for example, two-photon resonant four photon ionization spectrum overlaps with three-photon resonant four-photon ionization spectrum if there are high excited states simultaneously resonant with the two and three photons of one-color laser light. Such simultaneous resonances often occur for the high excited states of a molecule where the states are usually heavily congested.

Another way to reach a high excited state is to use two different lasers by stepwise excitations of a molecule via a well defined low lying excited state. A molecule is excited to a low lying excited state by the absorption of one-photon of laser frequency ω_1; then the excited molecule is further excited to a higher excited state by the one-photon absorption of another laser light of frequency ω_2. In this double resonance spectro-scopy, the high excited states reached from the well defined intermediate state by the ω_2 absorption are severely restricted by selection rules depending upon the nature of the selected intermediate state. As a result, the spectrum becomes simple, making the assignments of the high excited states easy. By selecting the intermediate state with the laser light of a fixed frequency ω_1, one can sort out the congested high excited states by the ω_2 absorption. Thus, the double resonance spectroscopy is most suitable for the elucidation of the energy level structure of a molecule in high energy region. This method also provides us with the dynamics of the intermediate state molecule by adjusting delay time between the ω_1 and ω_2 lasers.

In the present chapter, various kinds of two-color double resonance spectroscopies utilized for the detection of the high excited states of molecules are briefly described with examples mainly obtained in our laboratory.

2. Two-Color Laser Spectroscopies

In optical-optical double resonance spectroscopy, it is essentially important to excite a molecule to a well-defined low lying excited state with the first laser light of frequency ω_1. Since most of molecules exhibit well-resolved (ro)vibronic structures in the spectra due to the optically allowed electronic transition from the ground state X to the lowest excited state A in the vapor phase, it is favorable to choose a (ro)vibronic level in the A state as an intermediate excited state. The selective excitation to the (ro)vibronic level can be achieved by tuning the laser frequency ω_1 to a (ro)vibronic band in the A ← X transition which is associated with the selected (ro)vibronic level in the A state. In small molecules like diatomic molecules where individual (ro)vibronic bands are usually well resolved in the vapor spectrum, the selective excitation to a particular (ro)vibronic level is easily accomplished for the molecules in the vapor phase. However, at a high vapor pressure, the excited state molecules are greatly consumed by collisional relaxation and further excitation of the molecule to a high excited state with ω_2 becomes inefficient. Therefore, it is desirable to reduce the vapor pressure as low as possible. As a molecule becomes larger, the spectrum due to the A ← X transition becomes complex and complete resolution of the (ro)vibronic structure is very hard. Even when the (ro)vibronic structure is resolved, one has many chances in which more than two transitions originating from thermally populated ground-state levels accidentally coincide in energy. Therefore, even with the use of a narrow width ω_1, there is no guarantee for the selective excitation to a particular excited state.

To ensure the selective excitation, it is desirable to suppress the thermal population in the ground state as low as possible. The preparation of molecules populating only in the lowest rovibronic level in the ground state, that is, the molecules at absolute zero temperature, is ideal. Molecules approaching this ideal condition can be obtained by the use of the supersonic jet technique[3]. The expansion of a gaseous mixture of sample molecules and rare gas atoms (He or Ar) at several atm. into vacuum through a small nozzle orifice produces a supersonic jet, in which the molecule is cooled down by the collision-induced transfer of thermal energy of the molecule to translational energy of the rare gas atoms. The molecules in a supersonic jet are in a non-equilibrium state and the temperature of the molecules is generally different for the translational, rotational and vibrational degrees of freedom. In most cases, the translational and rotational temperatures are less than a few and 10 K, respectively. The vibrational temperature, however, is greatly different from molecule to molecule. In general, the molecule having many low frequency vibrational modes can be efficiently cooled down and its vibrational temperature becomes low. Therefore, in large polyatomic molecules, the molecules are virtually in the zero-point energy level in a supersonic jet, but, T_v for small molecules is fairly high. The fairly high vibrational temperature for small molecule does not impose serious problem for the selective excitation because the vibrational structure of the A ← X transition is generally sparse.

By the use of the supersonic jet technique, the excitation with ω_1 occurs practically from the vibrationless ground state and the preparation of the molecules in a specific vibronic level in the A state can be achieved. Since the excited molecules are collision-free in a supersonic jet, they have a fairly long life time and can be efficiently excited further to a higher excited state by the ω_2 absorption. Therefore, double-resonance spectroscopy applied to molecules in supersonic jets is most useful for the study of very high excited states of molecules. Various methods for the detection of the highly excited molecules produced by double resonance excitation will be described in following sections with examples.

2.1 Two-Color Multiphoton Ionization (MPI) Spectroscopy.

This is a most popular spectroscopic means for the observation of high excited states of a molecule and its principle is schematically shown in Fig. 1. A molecule in a jet is excited to a particular (ro)vibronic level in the A state by the ω_1 absorption. Then, the second laser light of tunable frequency ω_2 is introduced to induce the transition from the selected excited

8

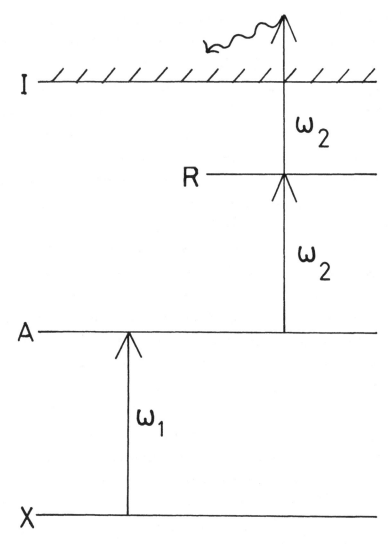

Fig. 1. Principle of two-color multiphoton ionization
spectroscopy. Wavy arrow shows the ion generation.

state to a higher excited state (R). The highly excited molecules absorb another ω_1 or ω_2 photon to reach the ionization
continuum of the molecule, thereby, the molecular ion is generated. Therefore, by detecting the ions while scanning the laser
frequency ω_2, one obtains the excitation spectrum corresponding
to the R ← A absorption spectrum. The method utilizes the fact
that the ions generated increase very much when ω_2 is resonant to
the R state. In total, the ionization process is three photon
(ω_1 + $2\omega_2$ or $2\omega_1$ + ω_2) ionization. Without resonant states,
the cross section of such a three-photon ionization is extremely
small. However, the cross section is greatly enhanced by the
simultaneous resonances at the two intermediate states A and R in
the multiphoton process. Two views are possible for the intermediate states. One is to regard them as virtual states in the
multiphoton process and the other is real excitations to the
intermediate states by sequence of the two independent one-photon
absorptions. The two processes may be distinguished by taking a
time delay between ω_1 and ω_2. The experimental results show that
in most cases the sequential excitations are predominant.

Two-color multiphoton ionization spectroscopy has first been
applied to I_2 vapor by Williamson and Compton[4]. They excited
the molecule to the B state by the first laser light of frequency
ω_1, then probed the D and F states lying in the 57000 ∿ 64000 cm^{-1}
region by the two-photon absorption of the second laser light
of ω_2. The ions were generated from the ionization continuum

which is reached by the absorption of one more photon of ω_2 from the D or F state. Ebata et al. measured the two-color double resonance enhanced four-photon ionization spectra of NO vapor[5,6]. The $A^2\Sigma^+$ state of NO was populated by the absorption of two photons of ω_1, and another photon ω_2 was used to reach the higher excited states (R) from the $A^2\Sigma^+$ state. By selecting various rovibronic levels in the $A^2\Sigma^+$ state by ω_1, the rovibronic level structures of $^2\Sigma^+$ and $^2\Pi$ states in the 65000 ∿ 70000 cm^{-1} region were clearly resolved. They also measured the double resonance MPI spectra by varying the delay time between ω_1 and ω_2 and obtained information on the collision-induced rotational relaxation in the $A^2\Sigma^+$, v = 0 and 1 states of NO[7]. It was shown that multiple quantum energy transfer induced by one collision occurs to at least ΔJ = 6. The two-color MPI spectra have also been reported in the study of the higher excited states of NO[8,9] and CO[10,11].

The two-color multiphoton spectroscopy is also useful for detecting the high excited states of a large polyatomic molecule lying below the adiabatic ionization potential. Fujii et al. observed the two-color MPI spectra of jet-cooled p-difluoro-benzene[12]. The jet-cooled molecule was excited to the zero-point level in the first excited singlet state $S_1(^1B_{2u})$. Then, the second laser light of frequency ω_2 was used to probe the higher excited states. They found four Rydberg series (of quan-

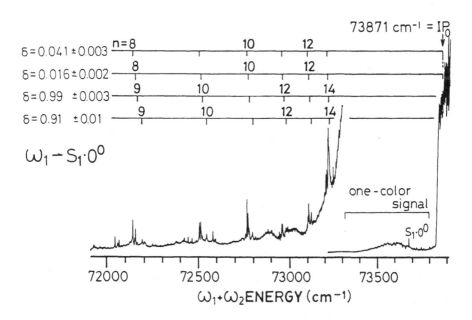

Fig. 2.   Two-color MPI spectrum of jet-cooled p-difluorobenzene obtained after exciting the molecule to the origin of S_1 with ω_1. The broad signal indicated is due to a complex formed between p-difluorobenzene and some impurity which appears with ω_2 alone. The sharp band at $73680 cm^{-1}$ is the 0^0_0 band of S_1 which appears also with ω_2 alone.   Adapted from ref.[12].

tum defect δ = 0.041, 0.016, 0.91 and 0.99) all converging to the
adiabatic ionization potential IP_0 at 73871 cm^{-1} (see Fig. 2).
From the observed results, they suggested a symmetry reduction in
the geometrical structure of the molecule in the Rydberg states
from the ground state symmetry D_{2h}.

2.2 Two-Color Ionization Threshold Spectra

In Fig. 1, when the one-photon transition from the A state
with ω_2 is resonant to the adiabatic ionization potential IP_0 of
a molecule, direct ionization from the A state to the ionization
continuum occurs and the ions are generated. By detecting the
ions while scanning the frequency ω_2, a threshold appears when ω_2
reaches IP_0. The threshold energy represents the zero-point
level in the electronic ground state of the cationic ion produced
by the ionization of the neutral molecule[13,14]. The threshold
is subject to a slight frequency shift from actual IP_0 due to the
electronic field which is imposed in the experiment for efficient
collection of the ions. The accurate IP_0 can be determined from
the convergence limit of a Rydberg series terminating to the
zero-point level of the ion if such a series is observed in the
MPI spectrum (as in the case shown in Fig. 2). The electric
field effect on ionization threshold has been treated by Cooke
and Gallagher by a classical model[15]. According to them, the
red shift of ionization threshold induced by electric field is
proportional to square root of the electric field. By assuming

the square root dependence, the actual IP_0 can be obtained from
the observed threshold energies at various electric field
strengths by extrapolating them to zero electric field. The
field effect is different from molecule to molecule. However, in
a typical case, the red shift induced by the electric field of
say, 30 v/cm, is about 20 cm^{-1}.

The intensity of threshold at IP_0 is determined by the
Franck-Condon factor between the vibrational level of a neutral
molecule in A state excited with ω_1 and the zero-point level of
the ground state ion apart from the electronic part of the transition moment. When geometrical structure is similar between the
neutral molecule in the A state and the ground state ion, the
Franck-Condon factor is very large for the transition with Δv =
0. Therefore, a strong IP_0 threshold can be observed when the
zero-point vibrational level in the A state is selected as an
intermediate state. Similarly, direct ionization preferentially
occurs from a vibrational level in the A state to the ionization
continuum belonging to the same vibrational level of the ion as
that in the A state. The ionization threshold observed represents vertical ionization potential IP_v corresponding to the
vibrational level of the ground state ion. The frequency difference IP_v - IP_0 gives the vibrational frequency of the ion.
Therefore, the measurement of the vertical ionization thresholds
by selecting various vibronic levels in the A state provides us
with the vibrational level structure of the ion. When the A

state is a Rydberg state, the vertical ionization threshold due to $\Delta v = 0$ transition appears strongly because the geometrical structure of a neutral molecule in Rydberg state is generally similar to that of the ion. Such a case was found for NO[16,17], diazabicyclooctane (DABCO)[18,19,20] and azabicyclooctane (ABCO)[21], in which A state is the 3s Rydberg state for all the molecules. Actually, for these molecules, the vibrational frequencies of the ion were found to be very similar to those in the A state.

When the potential and geometrical structure of a molecule in A state are considerably different from those of the ion, Franck-Condon factor becomes large also for the transitions with $\Delta v = 0$. Therefore, the two-color ionization spectrum exhibits several thresholds corresponding to the transitions with $\Delta v = 0$, 1, 2 ··· , and the intensity of each threshold reflects the Franck-Condon factor. Furthermore, a threshold corresponding to the transition from a selected vibrational level of a particular mode in the A state to a vibrational level of another mode in the ion occurs in the spectrum. When both modes belong to a same symmetry species, the appearance of the threshold implies the existence of Duschinsky rotation of the normal coordinates between the A state molecule and the ground-state ion. If the two modes are different in symmetry, the threshold suggests a difference in symmetry of geometrical structure between the A state molecule and the ion or vibronic coupling in these states.

Therefore, two-color ionization threshold spectrum provides useful information about the potential and geometrical structure of ion.

Two-color ionization threshold spectra are reported for many aromatic molecules, benzene[22], fluorobenzene[23,24], p-difluorobenzene[12], aniline[25], phenol[24,26], disubstituted benzenes[27], pyrazine[28], pyrimidine[29], 2-aminopyridine[30], naphthalene[22,31], indole[32,33] and 7-azaindole[26]. For all the molecules, the S_1 state serves as an intermediate A state. Since the S_1 states of these molecules are valence states (π, π^* or n, π^*), the geometrical structure and potential of the S_1 state molecule are considerably different from those of the ion. Therefore, in most cases, several thresholds appear in the spectrum. The vibrational level structure of the ion obtained from the observed thresholds is usually quite different from that of the S_1 state molecule, but rather close to that of the S_0 state molecule. For example, in pyrazine, most of the vibrational modes have similar frequencies between the S_0 state molecule and the ion, but quite different frequencies in the $S_1(n, \pi^*)$ state, as shown in Table I. The similarity and the difference are easily understood from the facts that the ground state ion of pyrazine is produced by the removal of an electron in the highest occupied non-bonding orbital of the ground state neutral molecule and that the S_1 state is created by the transfer of the non-bonding electron to the antibonding π^* orbital.

Table I. Comparison of vibrational frequencies (in cm^{-1}) among the S_0, S_1 and I_0 states of pyrazine.[a]

S_0 [b]	S_1 [b]	I_0 [c]	assignment
(0)	(30876)	(74915)	
919	383	509	$10a^1(b_{1g})$
832	467	895	$16b^2(b_{3u}xb_{3u})$
756	517	626	$5^1(b_{2g})$
596	583	631	$6a^1(a_g)$
1839	823	1080	$10a^2$
1517	945	1132	$10a^16a^1$
1014	966	1005	$1^1(a_g)$
	1033	1255	5^2
1231	1101	1251	$9a^1(a_g)$
1193	1167	1267	$6a^2$
	1243	1411	$4^2(b_{2g}xb_{2g})$
2752	1307	1712	$10a^3$
1578	1373	1508	$8a^1(a_g)$

a)   Data taken from ref.[28]

b)   The value was obtained from the study by Y. Udagawa, M. Ito and I. Suzuka, Chem. Phys. **46**, 237 (1980).

c)   Uncertainty of the value is ~ 5 cm^{-1}.

When the potential and geometrical structure are greatly different between a molecule in its A state and the ion, direct ionization does not give sharp threshold, but only a broad threshold is observed. Typical examples can be seen in van der Waals molecules and hydrogen-bonded complexes. Fig. 3 shows for example the two-color ionization threshold spectra of jet-cooled fluorobenzene and its van der Waals complex with CCl_4[23]. Both spectra were obtained from the excitations of the molecule and complex in their zero-point levels in the $S_1(\pi, \pi*)$ states. In the free molecule, a sharp ionization threshold appears at about 74200 cm^{-1}, while in the complex the ionization signal begins to appear at about 72000 cm^{-1} and gradually increases toward higher energy. The broad threshold observed for the complex indicates a great difference in the potential and geometrical structure between the S_1 state of the complex and its ion. In the S_1(or S_0) state, the van der Waals complex is formed by a weak van der Waals interaction. However, in the ion, a strong intermolecular interaction such as charge-induced dipole interaction will be added. Therefore, the geometrical structure of the complex ion will be greatly different from that of the neutral complex in the S_1 state. In such a case, Franck-Condon factor is negligibly small between the zero-point levels in the S_1 state and the ion, giving no threshold at IP_0. The direct ionization is possible only from the S_1 zero-point level to very high vibrational levels in the ion where the levels are heavily congested, resulting in a

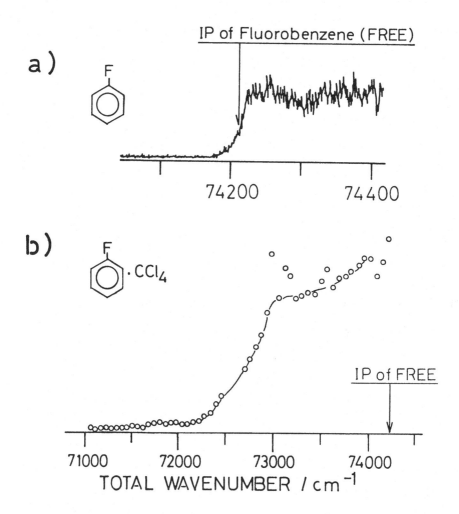

Fig. 3. Two-color ionization threshold spectra of a) fluoro-benzene and b) fluorobenzene-CCl_4 complex in a supersonic jet. Both spectra were obtained from the excitations of the fluorobenzene and the complex molecules in their 0^0 levels in the S_1 state. The electric field is 25 V/cm. Adapted from ref.[23].

broad threshold. A similar phenomenon was also found for the hydrogen-bonded complexes of phenol[24,26] and 7-azaindole[26], and the van der Waals complexes of pyrimidine[29], 2-aminopyridine[30], indole[32,33] and p-xylene[34].

Before closing this section, it should be mentioned that the two-color ionization threshold spectra of jet-cooled molecules usually give sharp thresholds, but sharp threshold does not appear for the molecules in the vapor phase. In the latter, direct transition to the ionization continuum can be assisted by molecular collisions existing in the vapor phase. The collision-induced ionization will be described in a later section.

2.3 <u>Two-Color</u> <u>Multiphoton</u> <u>Ionization</u> <u>Assisted</u> <u>by</u> <u>Autoionization</u>

There are many discrete bound states of a neutral molecule lying above the adiabatic ionization potential IP_0 of the molecule. These states can couple with the isoenergetic ionization continuum by a breakdown of the Born-Oppenheimer approximation and autoionize. Therefore, these states can be observed by two-color MPI assisted by autoionization which is schematically shown in Fig. 4. All sharp ionization peaks appearing above IP_0 in two-color MPI spectra can be assigned to these autoionizing levels of a neutral molecule. The autoionizing excited states so far observed by two-color MPI spectroscopy are all Rydberg states of molecules[35-44]. Since the geometrical structure of a molecule in a Rydberg state is in general very similar to that of the

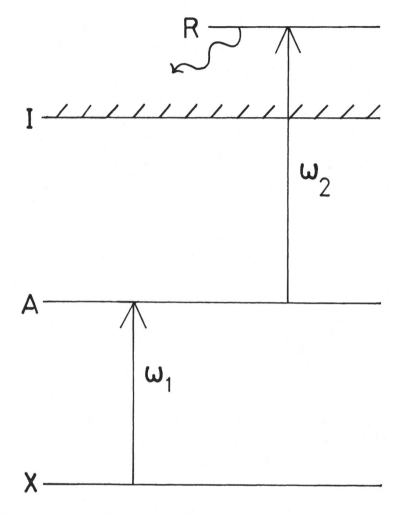

Fig. 4.   Principle of two-color multiphoton ionization assisted by autoionization.   Wavy arrow shows the generation of ion by autoionization from the R state.

ground state ion, vibrational or rotational autoionization is mainly responsible for the appearance of the Rydberg states.

Considering a particular vibrational level of a particular Rydberg state of a molecule, the wave function is approximately given by

$$\Phi_{Ryd} = \Psi_{Ryd}(n, \ell, m ; \mathtt{r}) \, H_{core}(\mathtt{r}_{N-1}, Q) \, \chi_{core}(Q)$$

where Ψ_{Ryd} is the wave function of the Rydberg electron and, H_{core} and χ_{core} are the electronic and vibrational wave functions, respectively, of the Rydberg core. $\mathtt{r}$ and Q stand for the coordinates of electrons and normal coordinates, respectively. On the other hand, the ionization continuum interacting with the Rydberg state is expressed by

$$\Phi_{ion + e^-} = \Psi_{e^-} \, H_{ion}(\mathtt{r}_{N-1}, Q) \, _{ion}(Q)$$

where Ψ_{e^-} is the wave function of a released electron and H_{ion} and χ_{ion} are the electronic and vibrational wave functions of the ion produced by the ejection of the electron from the neutral molecule. The matrix element for vibrational autoionization is given by

$$\langle \Phi_{ion+e^-} | \hat{H}'_{BO} | \Phi_{Ryd} \rangle$$
$$= \sum_i \frac{1}{M_i} \langle H_{ion} | \langle \Psi_{e^-} | \frac{\hbar}{i} \frac{\partial}{\partial Q_i} | \Psi_{Ryd} \rangle | H_{core} \rangle$$
$$\times \langle \chi_{ion} | | \frac{\hbar}{i} \frac{\partial}{\partial Q_i} | | \chi_{core} \rangle$$

where $\hat{H}_{BO}'$ represents the Born-Oppenheimer coupling operator.
For the Rydberg state, $\chi_{core} \approx \chi_{ion}$[45]. Assuming the harmonic
approximation for the vibrational wave function, the last term in
the above equation gives a selection rule of $\Delta v = \pm 1$. Taking
into account energy conservation, the selection rule for the
vibrational autoionization is concluded to be $\Delta v = - 1$.

This selection rule was actually confirmed for the large
polyatomic molecules such as aniline[46,47], p-difluorobenzene[12],
pyrazine[28], DABCO[19,48] and ABCO[21] in supersonic jets. As
an example, the experimental results obtained for jet-cooled
pyrazine are shown in Fig. 5. Fig. 5a is the two-color MPI
spectrum obtained after exciting jet-cooled pyrazine to the zero-
point level in $S_1(n,\pi^*)$ state with ω_1. A sharp threshold at
74900 cm^{-1} represents the adiabatic ionization potential IP_0. In
Fig. 5b is shown the two-color MPI spectrum obtained after
populating the molecule to the $10a^1(b_{1g})$ vibronic level in the S_1
state. A sharp threshold at 75400 cm^{-1} represents the vertical
ionization potential leading to the $10a^1$ vibrational level of the
ion ($I\ 10a^1$). In the spectrum, a Rydberg series of quantum
defect $\delta = 0.92$ appears which converges to the threshold at 75400
cm^{-1}. Therefore, the individual members of the Rydberg series
must be assigned to the $10a^1$ vibrational levels belonging to the
individual Rydberg states. Sudden appearance of the Rydberg
series in the energy region above IP_0 and its absence below IP_0
clearly indicate that the Rydberg series appears by autoioniza-

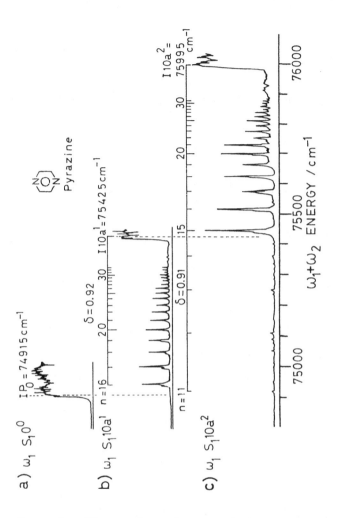

Fig. 5. Two-color MPI spectra of jet-cooled pyrazine due to the transition from a) the S_1 origin (30876 cm^{-1}), b) the $10a^1$ (b_{1g}, 383cm^{-1}) and c) the $10a^2$ (823 cm^{-1}) vibrational levels in the S_1 state. Calculated Rydberg series is shown by solid lines. The convergence limit is shown by vibrational state of the ion produced. Adapted from ref.[28].

tion. Fig. 5c shows the two-color MPI spectrum obtained after
exciting the molecule to the overtone level of the 10a mode
($10a^2$) in the S_1 state. A sharp threshold at 76000 cm^{-1} is due
to direct transition from the S_1 $10a^2$ level to the ionization
continuum belonging to the $10a^2$ level of the ion (I $10a^2$). The
Rydberg series of $\delta = 0.91$ terminating to the I $10a^2$ threshold is
assigned to the $10a^2$ vibrational levels of the Rydberg states.
This $10a^2$ Rydberg series appears only in the energy region above
the position where the $10a^1$ threshold (indicated by I $10a^1$ in the
figure) is situated, and the series suddenly disappears below
this position. The results clearly indicate that the $10a^2$
Rydberg state autoionizes to the continuum belonging to I $10a^1$,
but the autoionization to the continuum belonging to the zero-
point level of the ion, I 0^0, is prohibited. Therefore, there
exists a selection rule of $\Delta v = -1$ which is in exact accordance
with the selection rule for vibrational autoionization.

In transition from the A state to the ionization region, two
ionization processes simultaneously occur, one is direct transition from the A state to the ionization continuum described in
Section 2.2 and the other the ionization of the bound state molecule assisted by autoionization just mentioned above. Because of
these two ionization processes, the spectrum will become complex.
However, the spectra of large polyatomic molecules are often very
simple, and may be broadly classified into two cases. One typical case is seen in the two-color MPI spectra of pyrazine ob-

tained by the S_1 $10a^n$(n = 0, 1, 2, $10a(b_{1g})$) level excitation shown in Fig. 5. In the spectra, the $10a^n$ Rydberg states appear by the vibrational autoionization in the region between I $10a^n$ and I $10a^{n-1}$, where there exists no background ion signal arising from the direct transition to the ionization continuum. Another case is found in the two-color MPI spectra of the same molecule obtained by the S_1 $6a^n$(n = 1 and 2, $6a(a_g)$) level excitation, which are shown in Fig. 6. As seen from the figure, step functional ion signals appear with the thresholds at IP_0 and I $6a^1$ in the spectrum of the S_1 $6a^1$ level excitation and with three thresholds at IP_0, I $6a^1$ and I $6a^2$ in that of the S_1 $6a^2$ level excitation. These ion signals are due to the direct ionizations from the S_1 state to the ionization continua belonging to the $6a^n$ vibrational levels of the ion. In the same energy region, there must be the $6a^n$ Rydberg states, but they do not give apparent signals.

A sharp contrast between the two spectra can be explained by interference between the direct ionization and the autoionization via the Rydberg state from the S_1 state. Following Fano[49], the shape of a band due to the transition to a discrete level interacting with a continuum is determined by q defined by

$$q = M_d \ / \ VM_c$$

where V is the interaction matrix element between the discrete level and the continuum, and M_d and M_c are the transition moments to the discrete level and continuum, respectively, from a particular low lying level. In the case of the S_1 $10a^1$ level excitation, for example, q for the $10a^1$ Rydberg state is infinity because $M_c = 0$. $M_c = 0$ results from the fact that the transition from the S_1 $10a^1$ level to the ionization continuum belonging to the zero-point level of the ion, I $10a^0$, (the latter is interacting with the $10a^1$ Rydberg state) is symmetry forbidden because the 10a mode is of nontotally symmetry species of b_{1g} under the molecular symmetry D_{2h}. In such a case of $q = \infty$, no distortion of the band shape occurs, giving a sharp symmetric band. In the spectrum of the S_1 $6a^1$ level excitation, since the 6a mode is totally symmetric a_g species, the transition from the S_1 $6a^1$ level to the Rydberg $6a^1$ level as well as the direct transition to the continuum of I $6a^0$ are allowed, that is, $M_d \neq 0$ and $M_c \neq 0$. Therefore, q will be much smaller than infinity. When the interaction matrix element is large, q will have a small value. It is well known that the shape of a band becomes very asymmetric with a small q value, that is, the intensity of the band decreases on one wing side of the band and increases on the other side. Therefore, one cannot expect a sharp band, but a wavy signal extending over the interaction region will appear. Such wavy signals for the individual Rydberg states will overlap with each other, giving an irregular plateau on the direct ionization signal. The observed noisy plateau in the region between IP_0 and I $6a^1$ in Fig. 6 may be understood by the above explanation. The apparent absence of Rydberg series suggests that the

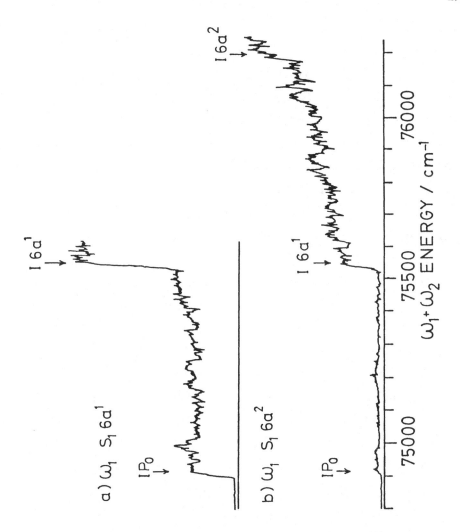

Fig. 6. Two-color MPI spectra of jet-cooled pyrazine obtained by a) the S_1 $6a^1(a_g$, 583 cm^{-1}) and b) S_1 $6a^2$(1167 cm^{-1}) level excitations. Adapted from ref.[28].

q value is very small, probably, less than 1. It is concluded that the appearance or apparent absence of the Rydberg series results from the optical activity of the interacting ionization continuum. Similar phenomenon was also found for jet-cooled aniline by Hager et al[47].

In all the polyatomic molecules so far studied by two-color MPI spectroscopy, the selection rule $\Delta v = -1$ for vibrational autoionization is strictly preserved. However, in NO, a violation of the selection rule was found for the autoionization, which might be ascribed to electronic autoionization[50-52].

2.4 Two-Color MPI Assisted by Collision

For an isolated molecule, the high excited states lying below the ionization potential IP_0 can not be detected by the two-color MPI method if the cross section for the ionization from the high excited state to the ionization continuum by the absorption of one more ω_1 or ω_2 photon is very small or if the laser power of ω_1 or ω_2 is very weak. In such a case, the two color MPI spectrum gives a sharp threshold at IP_0 but no signal appears in the region below IP_0. However, when a molecule is subject to collisions, the highly excited molecule pumped by the ω_2 absorption from the A state acquires the collisional energy and can be promoted to the ionization continuum. This situation is schematically shown in Fig. 7. An example of the collisional ionization is illustrated in Fig. 8 for DABCO[19,53]. Fig. 8b

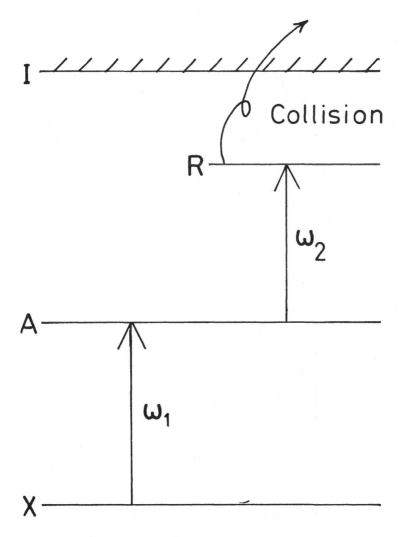

Fig. 7.   Principle of two-color multiphoton ionization assisted by collision.

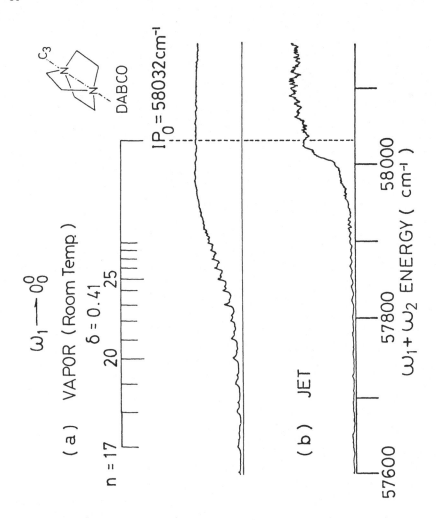

Fig. 8. Two-color MPI spectra obtained by exciting the molecule
to the S_1 origin (35783 cm^{-1}) by two-photon absorption;(a) vapor
at room temperature(0.25 Torr) and (b) supersonic jet (10 mm
downstream). The ionization potential IP_0 is shown by the broken
line. The electric field is 40 V/cm. Adapted from ref.[19].

shows the two-color MPI spectrum of DABCO in a supersonic jet
obtained by exciting the molecule to the S_1 origin (35783 cm^{-1})
by two-photon absorption. A sharp threshold slightly red-shifted
by the electric field effect is seen near IP_0 and no signal occur
in the region below IP_0. Fig. 8a shows the same two-color MPI
spectrum of DABCO vapor at room temperature. Now, the ion signal
begins to appear from the energy much lower than IP_0. Furthermore, a structure due to a Rydberg series is seen on the longest
wavelength tail. It is apparent that all these signals below IP_0
arise from the ionization of the high excited states lying below
IP_0 induced by molecular collisions existing in the vapor phase.
The energy difference between the onset of the ion signal and IP_0
reflects the maximum collisional energy. Therefore, adequate
utilization of molecular collision is useful for the detection of
high excited states lying below IP_0.

Under supersonic free-jet condition, the molecules are
supposed to be completely isolated and have no collision.
However, in high excited Rydberg states, the Rydberg electron is
moving far away from the core. For example, in hydrogen atom,
the average distance between the core and the (Rydberg) electron
amounts to as large as 700 Å for the principal quantum number n =
30. Therefore, the highly excited Rydberg molecule is regarded
as a huge molecule and collisional effects cannot be neglected
even under the supersonic jet condition. Fig. 9 shows for an
example the two-color MPI spectrum of jet-cooled trans-stilbene

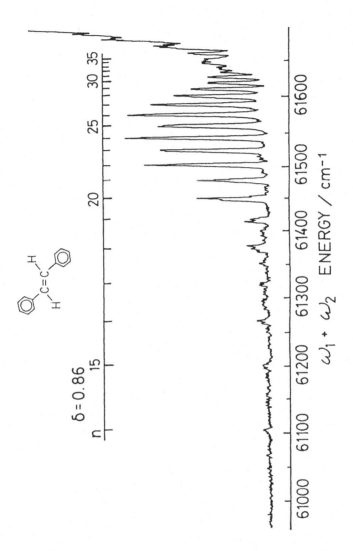

Fig. 9.   Two-color MPI spectrum of jet-cooled trans-stilbene measured with high gain. The ω_1 laser is fixed to the 0,0 band of the $S_1 \leftarrow S_0$ transition. The assignment of the Rydberg series is shown by solid lines. Adapted from ref.[54].

measured with a high gain[54]. The ω_1 laser frequency is fixed to the 0^0_0 band of the $S_1 \leftarrow S_0$ transition. Besides a strong and sharp threshold (IP_0) at 61700 cm^{-1}, a well-resolved Rydberg series converging to IP_0 appears. The Rydberg series is analyzed with the well-known Rydberg equation with $\delta = 0.86$ and $IP_0 =$ 61750 cm^{-1}. The appearance of the Rydberg series is apparently ascribed to the collisional ionization. It is seen from the figure that the intensity maximum is at $n \sim 25$ and the intensity decreases with larger and smaller n. This intensity distribution of the Rydberg series is easily explained by two factors; one is the collision probability which decreases with the decrease of n and the other is the transition probability from the S_1 state which decreases with the increase of n.

The Rydberg series observed by the two-color MPI assisted by collision is very useful for accurate determination of IP_0. Similar Rydberg series were observed also for NO[16,55] and p-difluorobenzene[12] in supersonic jets.

2.5 Two-Color Fluorescence Excitation Spectroscopy

In stead of monitoring ions, the high excited states of a molecule can be observed by monitoring the fluorescence originating from a high excited state (see Fig. 10). Two-color fluorescence excitation spectroscopy is very useful for small molecules like di-and tri-atomic molecules for which the fluorescence quantum yield for a high excited state is in general rather high.

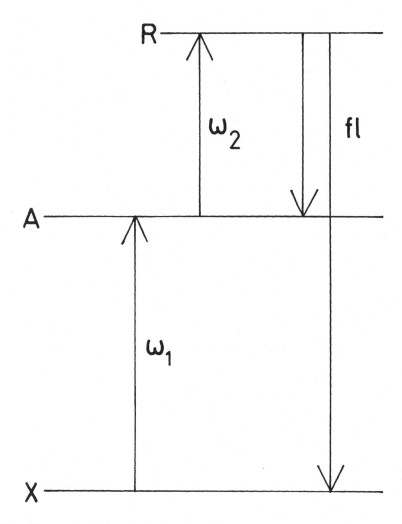

Fig. 10. Principle of two-color fluorescence excitation
spectroscopy. The fluorescence from the R state either to A or
to X state is monitored.

However, the application of this spectroscopy to large polyatomic molecules is severely restricted because of the extremely small fluorescence quantum yield. A typical example of this spectroscopy is found for halogen molecules, which have been extensively studied by Tanaka et al.[56-63]. Fig. 11 shows the energy level diagram of Br_2, for example. The molecule is excited to a particular rovibronic level in the $B^3\Pi(O_u^+)$ state with the first laser light of frequency ω_1. The excited molecules are excited further to higher excited states by the two photon absorption of $\omega_1 + \omega_2$ or $2\omega_2$. By detecting the ultraviolet fluorescence originating from the high excited state while scanning the frequency ω_2, one obtains the two-color fluorescence excitation spectrum in which the congested high excited states of the molecule are clearly resolved on rotational level. The most important achievement of the study is the findings of various new states including ion-pair states. Theoretically, it had been predicted that halogen molecule has many ion-pair states in which the molecule is dominated by ionic structure and the bond distance is anomalously large ($3 \sim 4$ A compared with 2.3 A in the ground state). Since Franck-Condon factor is very small for the direct transition from the ground state to such an ion-pair state, it is difficult to detect the ion-pair state by one-photon absorption spectroscopy. In double resonance spectroscopy, however, one can select a suitable intermediate state which has large Franck-Condon factors with both the ground state and the

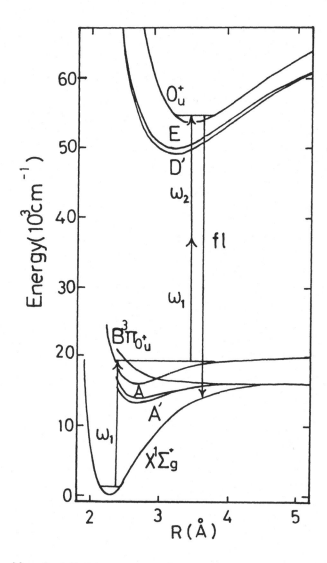

Fig. 11. Qualitative energy level diagram for the principle of the two-color double resonance experiment on Br_2.

ion-pair state (see Fig. 11). By this approach, many new ion-pair states have been found for various halogen molecules Cl_2[56-58], Br_2[59-62], I_2[63-69], IF[70], ICl[71-75] and IBr[76].

The application of two-color fluorescence excitation spectroscopy to the study of high excited states was also made for CO[77], alkaline dimers[78-92], CaF[93], BaF[94], BaO[95-98], CaCl[99], HgAr[100], NO_2[101], S_2O[102] and Cl_2CS[103].

2.6 Two-Color Fluorescence Dip Spectroscopy

As mentioned in a previous section, the method monitoring the fluorescence originating from high excited states is difficult to apply to large polyatomic molecules in which the fluorescence is usually quenched by various inter- and intramolecular processes. However, in many polyatomic molecules, the lowest electronic excited state $A(S_1)$ is fluorescent. Therefore, the two-color excitations of a molecule can be probed by selecting such a fluorescent state as an intermediate state and by monitoring the fluorescence from this intermediate state. The principle of this spectroscopy is shown in Fig. 12. A molecule in a jet is excited by the first laser light of frequency ω_1 to a selected level in the fluorescent A state, and then the second tunable laser light of frequency ω_2 is introduced to excite the molecule in the A state to higher excited states as usually done for the two-color experiments. In the presence of ω_1, the molecule in the A state emits the fluorescence of a constant intensity.

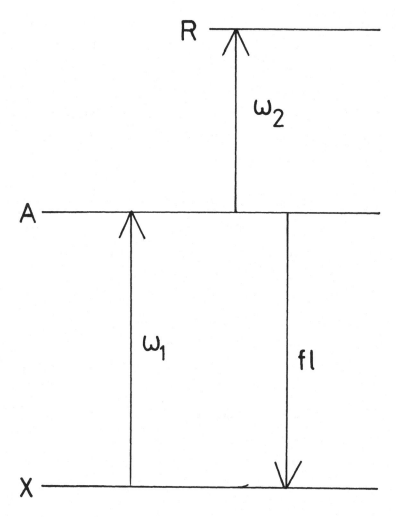

Fig. 12. Principle of two-color fluorescence dip spectroscopy.
The intensity of fluorescence from the A state exhibits
de-enhancement when ω_2 is resonant to the R state.

When the frequency ω_2 is resonant to a higher excited state R, the fluorescence intensity decreases because of the depopulation of the A state molecules caused by the R ← A absorption. Therefore, observing the total fluorescence while scanning the frequency of ω_2, one obtains so called two-color fluorescence dip spectrum. Two-color fluorescence dip spectroscopy is free from the ionization process of a molecule. Therefore, the method can be applied to higher excited states in a wide spectral region irrespective of the location of ionization potential. A good example illustrating this is shown for jet-cooled DABCO in Fig. 13a, which shows the two-color fluorescence dip spectrum of the molecule due to the transition from the S_1 20^1 vibronic level to the high energy region obtained by monitoring the total fluorescence from the S_1 state[48]. For comparison, the two-color MPI spectrum due to the same transition is shown in Fig. 13b. The MPI spectrum exhibits a sharp threshold of the vertical ionization IP_v leading to the 20^1 level of the ion and a Rydberg series converging to this threshold. The series appears only above the adiabatic ionization potential, indicating the appearance of the series by autoionization. In contrast to the MPI spectrum, the threshold IP_v does not occur in the fluorescence dip spectrum. Moreover, the Rydberg series appears strongly even below IP_0 and continues until as far as n = 4 which is located at 6000 cm^{-1} below IP_0. This example clearly shows a great advantage of the two-color fluorescence dip spectroscopy over MPI spectroscopy for

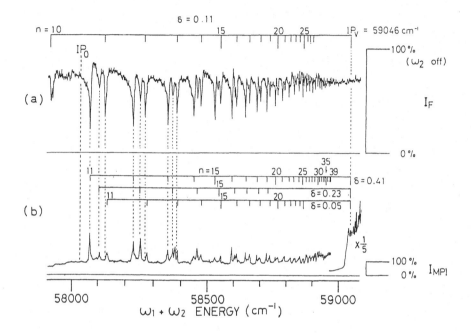

Fig. 13. Two-color fluorescence dip (a) and MPI (b) spectra of DABCO due to the transition from the S_1 20^1 (e', 1007 cm^{-1}) vibronic level for the energy region 58000-59000 cm^{-1}. Zero signal level (0%) and the signal level without ω_2 (100%) are indicated at the right side of each spectrum. The adiabatic ionization potential IP_0 and the vertical ionization potential IP_v are also shown. Calculated Rydberg series are shown by solid lines. Adapted from ref.[48].

the energy region to be covered.

Another advantage of the two-color fluorescence dip spectroscopy arises from the fact that the dip intensity directly reflects the cross section for the transition from the selected
fluorescent A level to a higher excited state reached by ω_2[6].
Assuming the steady-state condition in the A state, we obtain

$$A_{off}/A = \sigma I_2/k_f + 1$$

A and A_{off} are the fluorescence intensities (not dip intensity)
observed when the laser light of frequency ω_2 is and is not
resonant to the high excited state, respectively. k_f is the
fluorescence rate from the A state, I_2 the power of ω_2, and σ the
absorption cross section for the transition from the A state to
the high excited state. Therefore, A_{off}/A shows a linear dependence upon I_2. From the observed linear dependence, the absorption cross section σ can be obtained when k_f for the A state is
known. From the power dependence, the absorption cross section
was obtained for the transitions from the S_1 20^1 level to various
Rydberg states (n = 5 $\sim$39) involving 20^1 mode. The cross
section was found to decrease with the increase of n.

In the case where the high excited states of a molecule lying
above the adiabatic ionization potential are observed in both the
two-color fluorescence dip and MPI spectra, we may obtain information on the autoionization rates of the high excited states.
In this respect, it is interesting to see again the comparison

between the two-color fluorescence dip spectrum and the corresponding MPI spectrum in the region above IP_0 shown in Fig. 13. It looks as if the Rydberg series are identical between the two spectra. As shown in the figure, this is true for the series of δ = 0.41 and 0.23. However, the series of δ = 0.05 which appears in the fluorescence dip spectrum is absent in the MPI spectrum. In contrary, the series of δ = 0.11 appearing in the MPI spectrum is absent in the fluorescence dip spectrum. The result shows that the series of δ = 0.05 has an autoionization rate much lower than those of any other series. On the other hand, the series of δ = 0.11 has a large autoionization efficiency considering a small absorption cross section for the transition to this series as is shown from the fluorescence dip spectrum. A great difference in the autoionization efficiency for different Rydberg series of a molecule was also found for p-difluorobenzene[12].

The two-color fluorescence dip technique was successfully applied for the first time by Ebata et al. for NO[6,42,43]. They measured the two-color fluorescence dip spectra as well as the two-color MPI spectra of jet cooled NO due to the transitions from various rovibronic levels of the $A^2\Sigma^+$ state to the high Rydberg states (n = 6 $\sim$ 46). The rotational structures of these high Rydberg states were completely analyzed from the observed spectra. From the analyses, a gradual transition from Hund's case (b) to Hund's case (d) with the increase of n was experimentally proved for the first time. It was also found that the absorption cross section from the A state to the high Rydberg state is proportional to $1/n^3$ in agreement with the theoretical prediction[104]. The fluorescence dip spectroscopy has also been applied to the detection of the high excited states of NO[105] and CO[77].

Two-color fluorescence dip spectroscopy was also applied to the study of high excited valence states of molecules. Goto et al. observed the two-color fluorescence dip spectra of jet-cooled glyoxal obtained via the vibronic levels in the $S_1(n,\pi^*)$ state of the molecule[106]. Since the transition from the $S_1(n,\pi^*)$ state to the higher excited $S_n(n,\pi^*)$ state is orbital-allowed, the latter state is selectively observed in the fluorescence dip spectra. Thus, the highly excited $S_n(n,\pi^*)$ states were observed for the first time which are usually hidden by nearby strong π,π^* absorption in the one-photon transition from the ground state. Similarly, they also observed the higher excited n,π^* triplet states $T_n(n,\pi^*)$ of glyoxal by two-color phosphorescence dip spectroscopy in which the lowest excited triplet state $T_1(n,\pi^*)$ was populated by the first laser light of ω_1 and the $T_n(n,\pi^*) \leftarrow T_1(n,\pi^*)$ transition induced by the second laser light of ω_2 was probed by the dip in the total phosphorescence from the $T_1(n,\pi^*)$ state.

2.7 Two-Color Ionization Dip Spectroscopy

The principle of this spectroscopy is schematically shown in

Fig. 14. With the first laser light of frequency ω_1, one photon resonant two photon ionization occurs by resonance to a particular rovibronic level in A state. The ion signal generated is proportional to the population of the molecules in the A state and has a constant intensity with ω_1 alone. When the second laser light of tunable frequency ω_2 is introduced and it is in resonance with a higher excited state R, the population of the A state decreases by the ω_2 absorption from the A state to the R state; thereby the ion signal exhibits a dip at the resonance. The principle is same as that of two-color fluorescence dip spectroscopy, except that instead of the fluorescence the ion is monitored. This method can be applied to nonfluorescent molecules.

Two-color ionization dip spectroscopy was applied for the first time by Cooper and Wessel for I_2[107,108] and by Murakami et al. for toluene and aniline[109]. An example of two-color ionization dip spectrum is seen again for DABCO[48]. Fig. 15 shows the two-color ionization dip spectrum of jet-cooled DABCO which was obtained by monitoring the ion signal of the one color (ω_1) one-photon resonant two-photon ionization with the S_1 20^1 state as the one-photon resonant state. The observed ion dips are assigned to the 20^1 vibrational levels of three Rydberg states (δ= 0.41, 0.23 and 0.11) with n = 5 and 6 which are reached by the ω_2 absorption from the S_1 20^1 level. In the figure, the simultaneously measured two-color fluorescence dip spectrum is also given for comparison. It is seen that the

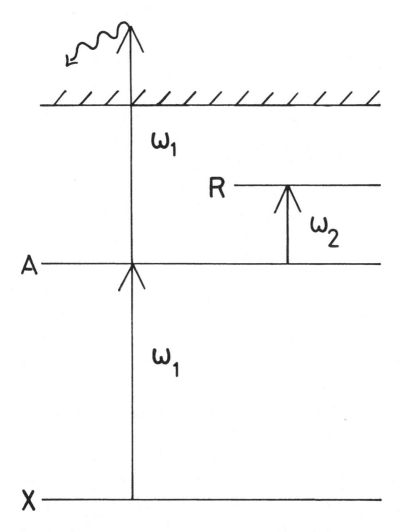

Fig. 14. Principle of two-color ionization dip spectroscopy. The ion signal resulting from the one-photon resonant two-photon ionization with ω_1 exhibits de-enhancement when ω_2 is resonant to the R state.

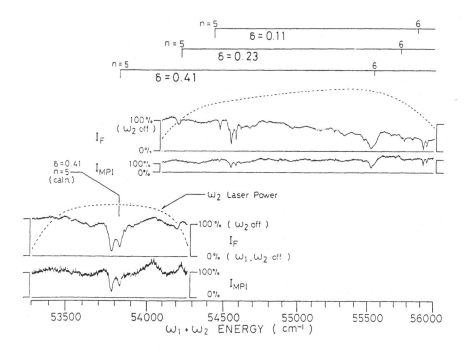

Fig. 15. Two-color ionization dip (lower) and fluorescence dip
(upper) spectra of jet-cooled DABCO due to the transition from
the 20^1 (e', $1007cm^{-1}$) vibrational level in the S_1 state. The
spectra (53300 $\sim$ 56000 cm^{-1}) are divided into two lasing region
of ω_2 laser dye (C-540A and R-6G). Relative laser power of ω_2 is
shown by a broken curve. Zero signal level (0 %) and the signal
level without ω_2 (100 %) are indicated at both side of each
divided spectrum. Calculated Rydberg series are indicated by
solid lines. Adapted from ref.[48].

spectrum is essentially the same between those obtained by the
two different methods. However, the quality of the spectrum
seems to be better in the fluorescence dip. This is because the
multiphoton ion signal varies sensitively with the ω_1 laser power
whose fine control is not so easy.

When R state lies above adiabatic ionization potential and
its autoionization rate is smaller than other nonradiative relaxation rate, an ion dip spectrum similar to that mentioned above
is observed with a high power of ω_2. Fig. 16 a and b show the
two-color fluorescence dip spectrum from the S_1 12^1 vibronic
level of jet-cooled ABCO and the simultaneously measured MPI
spectrum, respectively, taken under the condition of strong power
of ω_2[21]. The fluorescence dip spectrum consists of five
Rydberg series (δ= 0.09, 0.25, 0.39, 0.52 and 0.85), all of
which converge to IP_v = 62322 cm^{-1} corresponding to the 12^1
vibrational level of the ABCO ion. The corresponding MPI
spectrum exhibits de-enhancement of the ion signal at the
positions of the Rydberg bands. Thus, the observed MPI spectrum
looks very similar to the fluorescence dip spectrum except for
the appearance of sharp ionization thresholds at IP_0 and IP_v in
the former. When the ω_2 laser power is decreased, one obtains
the MPI spectrum shown in Fig. 16c. In this spectrum, the ion
signal is enhanced for the bands of the δ = 0.52 series, while
the bands of the δ = 0.09 and 0.39 series still show de-enhancement. When the laser power is further reduced, all the Rydberg

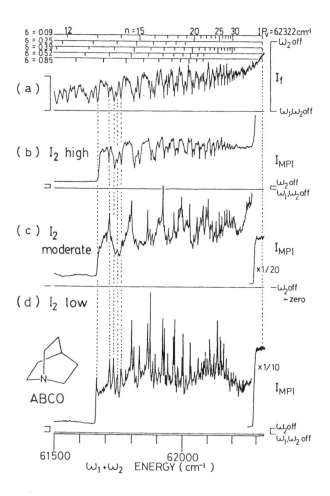

Fig. 16. Two-color fluorescence dip (a) and MPI (b-d) spectra of ABCO due to the transition from the 12^1 (a_1, 625 cm^{-1}) vibronic level in the S_1 state. The laser power of ω_2 is the strongest for (a) and (b), and is reduced in the order of (c) to (d). Adapted from ref.[21].

bands exhibit the ion signal enhancement as shown in Fig. 16d. The observed power dependence is approximately given based on rate equations by

$$\frac{M^+}{M^+_{off}} = (1 + \frac{\gamma_a}{\gamma_a+\gamma_n} \frac{\sigma_{III}}{\sigma_{II}})(\frac{k_f}{1+\sigma_{III}I_2})$$

where M^+ and M^+_{off} are numbers of the ions when ω_2 is and is not resonant to the R state, respectively. σ_{II} and σ_{III} are the absorption cross sections for the transitions to the ionization continuum and to the R state, respectively, from the A state (see Fig. 17). γ_a and γ_n are the rates of the autoionization and other non-radiative decay from the R state and k_f is the fluorescence rate of the A state. I_2 is the laser power of ω_2. The last term in the above equation is shown to be equal to the ratio of the fluorescence intensities (A/A_{off}) in resonance to off resonance to the R state in the corresponding fluorescence dip spectrum. Therefore, we have

$$\frac{M^+}{M^+_{off}} = (1 + \frac{\gamma_a}{\gamma_a+\gamma_n} \frac{\sigma_{III}}{\sigma_{II}}) \frac{A}{A_{off}}$$

This equation shows a linear relationship between M^+/M^+_{off} and A/A_{off}. The relationship was confirmed from the two-color MPI and fluorescence dip spectra obtained at various I_2 for the n = 16 band of the δ = 0.52 series. With reasonable estimation of σ_{II} and σ_{III}, the ratio γ_a/γ_n was obtained to be smaller than 10^{-2}. Therefore, the high Rydberg state of ABCO has a non-radiative decay channel much faster than the autoionization pro-

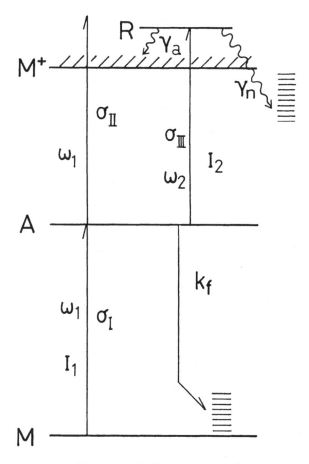

Fig. 17. Schematic diagram of fluorescence dip and ion dip by two-color excitation. Adapted from ref.[21].

cess. It is concluded that, when two-color MPI spectrum exhibits a dip at a Rydberg state lying above the adiabatic ionization potential, the Rydberg state has a fast non-radiative channel other than autoionization process. The fast non-radiative channel is probably predissociation channel.

In the energy region above the adiabatic ionization potential of a molecule, it happens that ion dip appears which is not associated with high Rydberg state. Instead of the upward transition with ω_2 from the A state, downward transition from the A state is possible which is caused by stimulated emission[107,109]. Suppose that the A state lies in energy above a half of the adiabatic ionization potential of a molecule, we have the ion signal due to the one-photon resonant two-photon ionization with ω_1 alone. When the second laser light of frequency ω_2 is resonant to the transition from the A state and a vibrational level in the X state, stimulated emission occurs and the population of the molecule in the A state decreases. The population decrease is reflected in the ion signal and gives an ion dip. Therefore, in such a case, the two-color ion dip observed is related with the ground state vibrational levels rather than high Rydberg states. An impressive example of this phenomenon was found by Suzuki et al. for trans-stilbene[54,110]. Fig. 18 shows the two-color MPI spectra obtained after exciting the jet-cooled trans-stilbene molecule to the 0^0 level of the S_1 state with ω_1. A sharp ionization threshold corresponds to the adiabatic ioniza-

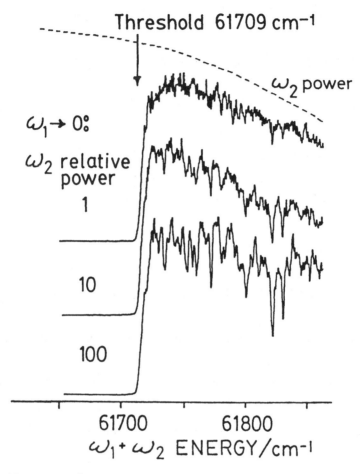

Fig. 18. Two-color MPI spectra of jet-cooled trans-stilbene.
The ω_1 laser is fixed to the 0,0 band of the $S_1 \leftarrow S_0$ transition
(32234 cm-1). The spectral heights of the three spectra obtained
with different ω_2 laser powers are normalized in the figure. The
dip intensity increases with increase of ω_2 laser power. ω_2
laser power 100 corresponds to a photon density of 5.6 x 10^{26}
photons cm^{-2}s^{-1}. Adapted from ref.[54].

tion by $\omega_1 + \omega_2$. In the energy region higher than the threshold, many sharp ion dips are found and their dip intensities increase with increase of the ω_2 laser power as seen in the figure. The appearance of the ion dip is not limited to the region near the threshold, but can be seen in the entire spectral region from the threshold to $\omega_1 + \omega_2 = 64500$ cm^{-1} (the shortest wavelength limit of ω_2 available). It was found that the observed ion dips exactly coincide in energy measured by $\omega_1 - \omega_2$ with the ground state vibrational energies obtained from the dispersed fluorescence spectrum from the S_1 state.

As seen from the above examples, ion dips observed in two-color MPI spectrum are due to a large non-radiative relaxation process of high excited state in some cases and to stimulated emission from the intermediate state in other cases. Therefore, we should be careful in the assignment of the observed dips. The two-color ion dip spectrum is also reported for NH_3[111].

2.8 Population Labeling Spectroscopy

The two-color double resonance spectroscopies described in previous sections have two advantages. One is that the high excited states of a molecule can be reached with two or more photons of low energies by the use of stepwise excitation via an intermediate state. The other is that the observed spectrum is simple because the transitions from the selected intermediate state is severely restricted by selection rules. However, the

high excited states detected by the sequential excitations are
sometimes different from those observed in the usual one-photon
absorption spectroscopy. For example, in the molecule having
inversion symmetry, only the high excited states belonging to g
species are observed in the two photon absorption via an intermediate state (u) from the ground state (g) molecule. On the
other hand, the states belonging to u species are allowed in the
one-photon absorption from the ground state. Therefore, the
direct one-photon absorption from the ground state is complementary to the stepwise two photon absorption on which all the two-
color double resonance spectroscopies described before are based.
In order to obtain the whole picture of the high excited states
of a molecule, the direct one-photon absorption is essentially
important. In this sense, the vacuum ultraviolet absorption
spectrum of a molecule in the vapor phase provides us with
information of the high excited states of a molecule which is
never obtained from other means. The vapor spectrum is usually
congested by the bands due to the transitions from various
thermally populated rovibronic level in the ground state. One
way to remove the spectral congestion is to prepare the jet-
cooled molecules. Then, the transitions only from the zero-point
level in the ground state remain and the spectrum becomes simple.
However, at the same time, we lose the information on the transitions from the thermally populated ground state. Population
labeling spectroscopy described in this section provides a way to

make the spectrum simple without losing the information.

The principle of population labeling spectroscopy is shown in Fig. 19. The first laser light of a fixed frequency ω_1 is tuned to the well defined A ← X transition of a molecule in the vapor phase which is associated with a particular (ro)vibronic level in the ground state, and the fluorescence from the A state is monitored. Then, the second laser light of tunable frequency ω_2 is scanned through the R ← X absorption region. When the frequency ω_2 coincides with that of the R ← X transition involving the same rovibronic level in the ground state as that for the A ← X transition, the population of this particular rovibronic level in the ground state decreases; thereby the fluorescence from the A state decreases in intensity. In this way, we can selectively observe the R ← X transitions originating from the particular rovibronic level in the ground state. By changing the selected rovibronic level, the congested band system of the R ← X absorption can be classified into many simple spectra each of which represents the transitions from the particular level selected. Since the selected rovibronic level is well defined from the analysis of the A ← X absorption spectrum, the assignment of the high excited state reached by the transition from the selected level can be easily made. The population of the ground-state rovibronic level can be monitored not only by the fluorescence from the A state but also by the multiphoton ion signal via the A state, the R ← X modulated absorption, or the

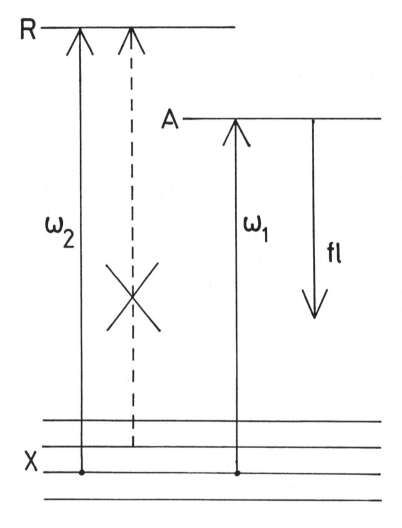

Fig. 19. Principle of population labeling spectroscopy. The fluorescence from the A state which is reached by ω_1 absorption from a particular level in the X state is monitored. The fluorescence intensity decreases when ω_2 is resonant to the transition from the particular level in the X state to the R state.

polarization rotation of ω_1.

The population labeling spectroscopy has been first applied by Kaminsky et al. for Na_2 molecule[112]. They obtained the modulated population spectrum due to the $A^1\Sigma_u \leftarrow X$ transition by labeling the $B^1\Pi_u - X\ ^1\Sigma_g^+$ state. Johnson et al. observed the population labeling spectrum of jet-cooled CO_2^+ and analyzed the $B^2\Sigma_u^+(000) \leftarrow X^2\Pi_{g3/2}(000)$ transition[113]. They used the $A^2\Pi_{u3/2} \leftarrow X^2\Pi_{g3/2}$ transition for labeling the ground state level and the population decrease was monitored by the fluorescence from the A state. The population labeling spectroscopy has also been applied to NO_2[114], Cs_2[115], NaCs[116], BaI[117,118], N_2[119] and O_2^+[120,121], but never to large polyatomic molecules. A great disadvantage of this method for the study of the high excited state of a molecule is that it requires vacuum ultra-violet high energy photon. Recent advance in vacuum ultraviolet laser will soon make this spectroscopy a powerful means for the study of high excited states of molecules.

3.   Conclusion

Various types of two-color double resonance spectroscopies which can be utilized for the study of high excited states of molecules were briefly described with examples. The two-color double resonance spectroscopies are complementary to the traditional vacuum ultraviolet absorption spectroscopy. However, the latter spectroscopy needs high power vacuum ultraviolet laser or synchrotron radiation which are still not popular in a small laboratory scale. The great advantage of the two-color double resoance spectroscopies is that the high excited states of molecules can be studied with low energy photons for which the usual spectroscopic apparatus for visible or ultraviolet region can be used without any modification. In addition to this experimental feasibility, the congested structure of the high excited states of a molecule can be resolved into simplified structures by the selection of the intermediate A state which is generally well studied. The simplification brought about by the well-defined intermediate state leads to unambiguous assignments of the high excited states. The two-color double resonance spectroscopy also opens a way to high resolution in the high energy region which was hitherto limited by the resolution of a vacuum ultraviolet spectrometer. It is expected from these advantages that the two-color double resonance spectroscopies will be widely used for the study of highly excited molecules and ions.

Reference

1] M. B. Robin, Higher excited states of polyatomic molecules Vol. I, II, III (Academic Press, New York 1974, 1975, 1985).

2] P. M. Johnson and C. E. Otis, Ann. Rev. Phys. Chem. 32, 139 (1981).

3] D. H. Levy, Ann. Rev. Phys. Chem. 31, 197 (1980).

4] A. D. Williamson, and R. N. Compton, Chem. Phys. Lett. 62, 295 (1979).

5] T. Ebata, T. Imajo, N. Mikami, and M. Ito, Chem. Phys. Lett. 89, 45 (1982).

6] T. Ebata, N. Mikami, and M. Ito, J. Chem. Phys. 78, 1132 (1983).

7] T. Ebata, Y. Anezaki, M. Fujii, N. Mikami, and M. Ito, Chem. Phys. 84, 151 (1984).

8] W. Y. Cheung, W. A. Chupka, S. D. Colson, D. Gauyacq, P. Avouris, and J. J. Wynne, J. Chem. Phys., 78, 3625 (1983).

9] W. Y. Cheung, W. A. Chupka, S. D. Colson, D. Gauyacq, P. Avouris, and J. J. Wynne, J. Phys. Chem. 90, 1086 (1986).

10] W. R. Ferrell, C. H. Chen, M. G. Payne, and R. D. Willis, Chem. Phys. Lett. 97, 460 (1983).

11] G. Sha, X. Zhong, S. Zhao, and C. Zhang, Chem. Phys. Lett. 110, 405 (1984).

12] M. Fujii, T. Kakinuma, N. Mikami, and M. Ito, Chem. Phys. Lett. 127, 297 (1986).

13] S. Leutwyler, M. Hofmann, H. P. Harri, and E. Schumacher, Chem. Phys. Lett. 77, 257 (1981).

14] D. Eisel, and W. Demtroeder, Chem. Phys. Lett. 88, 481 (1982).

15] W. E. Cooke, and T. F. Gallagher, Phys. Rev. A. 17, 1226 (1978).

16] T. Ebata, Y. Anezaki, M. Fujii, N. Mikami, and M. Ito, J. Phys. Chem. 87, 4733 (1983).

17] K. Mueller-Dethlefts, M. Sander, and E. W. Schlag, Chem. Phys. Lett. 112, 291 (1984).

18] D. H. Parker, and M. A. El-Sayed, Chem. Phys. 42, 379 (1979).

19] M. Fujii, T. Ebata, N. Mikami, and M. Ito, Chem. Phys. Lett. 101, 578 (1983).

20] M. A. Smith, J. W. Hager, and S. C. Wallace, J. Phys. Chem. 88, 2250 (1984).

21] M. Fujii, N. Mikami, and M. Ito, Chem. Phys. 99, 193 (1985).

22] M. A. Duncan, T. G. Dietz, and R. E. Smalley, J. Chem. Phys. 75, 2118 (1981).

23] N. Gonohe, A. Shimizu, H. Abe, N. Mikami, and M. Ito, Chem. Phys. Lett. 107, 22 (1984).

24] N. Gonohe, H. Abe, N. Mikami, and M. Ito, J. Phys. Chem. 89, 3642 (1985).

25] M. A. Smith, J. W. Hager, and S. C. Wallace, J. Chem. Phys. 80, 3097 (1984).

26] K. Fuke, H. Yoshiuchi, and K. Kaya, Chem. Phys. Lett. 108, 179 (1984).

27] A. Oikawa, H. Abe, N. Mikami, and M. Ito, Chem. Phys. Lett. 116, 50 (1985).

28] A. Goto, M. Fujii, and M. Ito, J. Phys. Chem in press.

29] N. Mikami, Y. Sugahara, and M. Ito, J. Phys. Chem. 90, 2080 (1986).

30] J. Hager, and S. C. Wallace, J. Phys. Chem. 89, 3833 (1985).

31] D. E. Cooper, R. P. Frueholz, C. M. Klimcak, and J. E. Wessel, J. Phys. Chem. 86, 4892 (1982).

32] J. Hager, M. Ivanco, M. A. Smith, and S. C. Wallace, Chem. Phys. Lett. 113, 503 (1985).

33] J. Hager, M. Ivanco, M. A. Smith, and S. C. Wallace, Chem. Phys. 105, 397 (1986).

34] P. D. Dao, S. Morgan, and A. W. Castleman Jr., Chem. Phys. Lett. 113, 219 (1985).

35] N. Bjerre, R. Kachru, and H. Helm, Phys. Rev. A. 31, 1206 (1985).

36] M. W. McGeoch, and R. E. Schlier, Chem. Phys. Lett. 99, 347 (1983).

37] D. Eisel, W. Demtroeder, W. Muller, and P. Botschwina, Chem. Phys. 80, 329 (1983).

38] S. Martin, J. Chevaleyre, S. Valignat, J. P. Perrot, M. Broyer, B. Cabaud, and A. Hoareau, Chem. Phys. Lett. 87, 235 (1982).

39] S. Leutwyler, T. Heinis, and M. Jungen, H. P. Harri, and E. Schumacher, J. Chem. Phys. 76, 4290 (1982).

40] M. C. Bordas, M. Broyer, J. Chevaleyre, P. Labastie, and S. Martin, J. Phys. $\underline{46}$, 27 (1985).

41] M. Broyer, J. Chevaleyre, G. Delacretaz, S. Martin, and L. Woeste, Chem. Phys. Lett. $\underline{99}$, 206 (1983).

42] Y. Anezaki, T. Ebata, N. Mikami, and M. Ito, Chem. Phys. $\underline{89}$ 103 (1984).

43] Y. Anezaki, T. Ebata, N. Mikami, and M. Ito, Chem. Phys. $\underline{97}$ 153 (1985).

44] E. E. Eyler, W. A. Chupka, S. D. Colson, and D. T. Biernacki, Chem. Phys. Lett. $\underline{119}$, 117 (1985).

45] R. S. Berry, J. Chem. Phys. $\underline{45}$, 1228 (1966).

46] J. Hager, M. A. Smith, and S. C. Wallace, J. Chem. Phys. $\underline{83}$, 4820 (1985).

47] J. Hager, M. A. Smith, and S. C. Wallace, J. Chem. Phys. $\underline{84}$, 6771 (1986).

48] M. Fujii, T. Ebata, N. Mikami, and M. Ito, J. Phys. Chem. $\underline{88}$, 4265 (1984).

49] U. Fano, Phys. Rev. $\underline{124}$, 1866 (1961).

50] Y. Achiba, K. Sato, and K. Kimura, J. Chem. Phys. $\underline{82}$, 3959 (1985).

51] J. C. Miller, R. N. Compton, J. Chem. Phys. $\underline{84}$, 675 (1986).

52] J. W. J. Verschuur, J. Kimman, H. B. Van Linden Van den Heuvell, and M. J. Van der Wiel, Chem. Phys. $\underline{103}$, 359 (1986).

53] G. J. Fisanick, T. S. Elchelberger IV, M. B. Robin, and N. A. Kuebler, J. Phys. Chem. $\underline{87}$, 2240 (1983).

54] T. Suzuki, N. Mikami, and M. Ito, Chem. Phys. Lett. <u>120</u>, 333 (1985).

55] M. Seaver, W. A. Chupka, S. D. Colson, and D. Gauyacq, J. Phys. Chem. <u>87</u>, 2226 (1983).

56] T. Ishiwata, I. Fujiwara, T. Shinzawa, and I. Tanaka, J. Chem. Phys. <u>79</u>, 4779 (1983).

57] T. Ishiwata, T. Shinzawa, T. Kusayanagi, and I. Tanaka, J. Chem. Phys. <u>82</u>, 178B (1985).

58] T. Shinzawa, A. Tokunaga, T. Ishiwata, I. Tanaka, J. Chem. Phys. <u>83</u>, 5407 (1985).

59] T. Ishiwata, T. Shinzawa, A. Tokunaga, and I. Tanaka, Chem. Phys. Lett. <u>101</u>, 350 (1983).

60] T. Shinzawa, A. Tokunaga, T. Ishiwata, and I. Tanaka, J. Chem. Phys. <u>80</u>, 5909 (1984).

61] T. Ishiwata, A. Tokunaga, T. Shinzawa, I. Tanaka, Bull. Chem. Soc. Jpn. <u>57</u>, 1317 (1984).

62] T. Ishiwata, A. Tokunaga, T. Shinzawa, and I. Tanaka, Bull. Chem. Soc. Jpn. <u>57</u>, 1469 (1984).

63] T. Ishiwata, A. Tokunaga, T. Shinzawa, and I. Tanaka, J. Mol. Spectrosc. <u>117</u>, 89 (1986).

64] U. Heemann, H. Knockel, and E. Tiemann, Chem. Phys. Lett. <u>90</u>, 17 (1982).

65] J. Chevaleyre, J. P. Perrot, J. M. Chastan, S. Valignat, and M. Broyer, Chem. Phys. <u>67</u>, 59 (1982).

64

66] J. B. Koffend, R. Bacis, M. Broyer, J. P. Pique, and S. Churassy, Laser Chem. $\underline{1}$, 343 (1983).

67] A. J. Bouvier, R. Bacis, A. Bouvier, M. Broyer, S. Churassy, and J. P. Perrot, Opt. Commun. $\underline{51}$, 403 (1984).

68] J. P. Perrot, A. J. Bouvier, A. Bouvier, B. Femelat, and J. Chevaleyre, J. Mol. Spectrosc. $\underline{114}$ 60 (1985).

69] K. Kasatani, J. Ito, M. Kawasaki, and H. Sato, Chem. Phys. Lett. $\underline{122}$, 113 (1985).

70] B. K. Clark, and I. M. Littlewood, Chem. Phys. $\underline{107}$, 97 (1986).

71] J. C. D. Brand, U. D. Deshpande, A. R. Hoy, and E. J. Woods, Can. J. Chem. $\underline{61}$, 846 (1983).

72] J. C. D. Brand, D. Bussieres, A. R. Hoy, S. M. Jaywant, Can. J. Phys. $\underline{62}$, 1947 (1984).

73] J. C. D. Brand, D. Bussieres, A. R. Hoy, and S. M. Jaywant, Chem. Phys. Lett. $\underline{109}$, 101 (1984).

74] D. Bussieres, and A. R. Hoy, Can. J. Phys. $\underline{62}$, 1941 (1984).

75] J. C. D. Brand, A. R. Hoy, and S. W. Jaywant, J. Mol. Spectrosc. $\underline{106}$, 388 (1984).

76] J. C. D. Brand, A. R. Hoy, and C. A. Risbud, J. Mol. Spectrosc. $\underline{113}$, 47 (1985).

77] P. Klopotek, and C. R. Vidal, J. Opt. Soc. Am. B : Opt. Phys. $\underline{2}$, 869 (1985).

78] R. A. Bernheim, L. P. Gold, P. B. Kelly, C. Kittrell, and D. K. Veirs, Phys. Rev. Lett. $\underline{43}$, 123 (1979).

79] R. A. Bernheim, L. P. Gold, P. B. Kelly, C. Kittrell, and D. K. Veirs, Vhem. Phys. Lett. 15, 104 (1980).

80] R. A. Bernheim, L. P. Gold, P. B. Kelly, T. Tipton, and D. K. Veirs, J. Chem. Phys. 74, 2749 (1981).

81] R. A. Bernheim. L. P. Gold, P. B. Kelly, C. Tomczyk, and D. K. Veirs, J. Chem. Phys. 76, 3249 (1981).

82] R. A. Bernheim, L. P. Gold, and T. Tipton, Chem. Phys. Lett. 92, 13 (1982).

83] R. A. Bernheim, L. P. Gold, P. B. Kelly, T. Tipton, and D. K. Veirs, J. Chem. Phys. 76, 57 (1982).

84] R. A. Bernheim, L. P. Gold, and T. Tipton, J. Chem. Phys., 78, 3635 (1983).

85] R. A. Bernheim, L. P. Gold, T. Tipton, and D. D. Konowalow, Chem. Phys. Lett. 105, 201 (1984).

86] X. Xie, and R. W. Field, Chem. Phys. 99, 337 (1985).

87] X. Xie, and R. W. Field, J. Chem. Phys. 83, 6193 (1985).

88] B. Hemmerling, S. B. Rai, and W. Demtroeder, Z. Phys. A 320, 135 (1985).

89] S. B. Rai, B. Hemmerling, and W. Demtroeder, Chem. Phys. 97, 127 (1985).

90] L. Li, and R. W. Field, J. Phys. Chem. 87, 3020 (1983).

91] L. Li, S. F. Rice, and R. W. Field, J. Chem. Phys. 82, 1178 (1985).

92] L. Li, and R. W. Field, J. Mol. Spectrosc. 117, 245 (1986).

93] P. F. Bernath and R. W. Field, J. Mol. Spectrosc. 82, 339 (1980).

94] P. C. F. Ip, P. F. Bernath, and R. W. Field, J. Mol. Spectrosc. 89, 53 (1981).

95] R. W. Field, G. A. Cappell, and M. A. Revelli, J. Chem. Phys. 63, 3228 (1975).

96] R. A. Gottscho, J. B. Koffend, R. W. Field, and J. R. Lombardi, J. Chem. Phys. 68, 4110 (1978).

97] R. A. Gottscho, J. Chem. Phys. 70, 3554 (1979).

98] R. A. Gottscho, J. B. Koffend, and R. W. Field, J. Mol. Spectrosc. 82, 310 (1980).

99] P. J. Domaille, T. C. Steimle, and D. O. Harris, J. Chem. Phys. 68, 4977 (1978).

100] W. H. Breckenridge, M. C. Dunal, C. Jouvet, and B. Soep, Chem. Phys. Lett. 122, 181 (1985).

101] K. Tsukiyama, K. Shibuya, K. Obi, and I. Tanaka, J. Chem. Phys. 82, 1147 (1985).

102] K. Tsukiyama, D. Kobayashi, K. Obi, and I. Tanaka, Chem. Phys. 84, 337 (1984).

103] R. N. Dixon, and C. M. Western, J. Mol. Spectrosc. 115, 74 (1986).

104] J. W. C. Johns, Molecular spectroscopy, Vol.2 (The chemical Society, London, (1974) p,513

105] M. R. Taherian, P. C. Coshy, and T. G. Slanger, J. Chem. Phys. 83, 3878 (1985)

106] A. Goto, M. Fujii, N. Mikami, and M. Ito, Chem. Phys. Lett. 119, 17 (1985).

107] D. E. Cooper, C. M. Klimcak, and J. E. Wessel, Phys. Rev. Lett. 46, 324 (1981).

108] D. E. Cooper, and J. E. Wessel, J. Chem. Phys. 76, 2155 (1982).

109] J. Murakami, and K. kaya, and M. Ito, Chem, Phys, Lett. 91, 401 (1982).

110] T. Suzuki, N. Mikami, and M. Ito, J. Phys. Chem. 90, 6431 (1986).

111] J. Xie, G. Sha, X. Zhang, and C. Zhang, Chem. Phys. Lett. 124, 99 (1986).

112] M. E. Kaminsky, R. T. Hawkins, F. V. Kowalski, and A. L. Schawlow, Phys. Rev. Lett. 36, 671 (1976).

113] M. A. Johnson, J. Rostas, and R. N. Zare, Chem. Phys. Lett. 92, 225 (1982).

114] M. Raab, H. J. vedder, and D. Zevgolis, J. Mol. Struct 59, 291 (1980).

115] M. Raab, G. Hoening, W. Demtroeder, and C. R. Vidal, J. Chem. Phys. 76, 4370 (1982).

116] U. Diemer, H. Weickenmeier, M. Wahl, and W. Demtroeder, Chem. Phys. Lett. 104, 489 (1984).

117] M. A. Johnson, C. R. Webster, and R. N. Zare, J. Chem. Phys. 75, 5575 (1981).

118] M. A. Johnson, and R. N. Zare, J. Chem. Phys. $\underline{82}$, 4449 (1985).

119] K. Miyazaki, H. Scheingraber, and C. R. Vidal, Phys. Rev. Lett. $\underline{50}$, 1046 (1983).

120] P. C. Cosby, and H. Helm, J. Chem. Phys. $\underline{76}$, 4720 (1982).

121] N. Bjerre, T. Andersen, M. Kaivola, and O. Poulsen, Phys. Rev. $\underline{31}$, 167 (1985).

RESONANTLY ENHANCED MULTIPHOTON IONIZATION -- PHOTOELECTRON SPECTROSCOPY AS A PROBE OF MOLECULAR PHOTOPHYSICS AND PHOTOCHEMISTRY[*]

S. T. Pratt, P. M. Dehmer, and J. L. Dehmer

Argonne National Laboratory, Argonne, Illinois 60439

[*]Work supported by the U.S. Department of Energy, Office of Health and Environmental Research, under Contract W-31-109-Eng-38.

1. INTRODUCTION

For the last twenty years, single photon ultraviolet photoelectron spectroscopy (UV-PES) has been a powerful technique for the study of the electronic structure of molecular ions and for the determination of binding energies of neutral molecular orbitals.[1-6] The combination of photoelectron spectroscopy with broadly tunable synchrotron light sources has enabled the determination of photoionization branching ratios and photoelectron angular distributions as a function of wavelength.[7] The interplay between this experimental work and the theoretical calculations that were stimulated by it has produced rapid advances in the understanding of molecular photoionization dynamics.

In recent years, the development of resonantly enhanced multiphoton ionization-photoelectron spectroscopy (REMPI-PES) has resulted in an enormous increase in the scope and variety of problems that can be addressed by PES.[8-10] As in UV-PES, REMPI-PES relies on the determination of the kinetic energies and angular distributions of electrons produced by photoionization; the essential difference between the two techniques arises from differences in the ionization process. In UV-PES, a single photon is used to ionize a (typically) thermal distribution of vibrational and rotational levels of the ground electronic state of the target molecule. In general, there is no control over which vibrational modes of an ionic state will be excited, as these modes are determined by symmetry considerations and by Franck-Condon factors between the ground state and the ionic state. In addition, it is usually quite difficult to obtain UV-PES spectra of minor components of mixtures, such as free radicals or van der Waals molecules, owing to interference from photoelectrons ejected from the major components of the mixture. Although UV-PES spectra of a number of free radicals[11-13] and van der Waals molecules[14-17] have been obtained, usually great effort is required to overcome these difficulties.

In REMPI-PES, a resonant multiphoton process is used to ionize the target molecule. In a typical (m+n) REMPI-PES study, an excited state of the neutral molecule is prepared by m-photon excitation (usually m=1-3) using radiation from a pump laser; this excited state is subsequently ionized by n-photon excitation (usually n=1) using radiation from either the pump laser or from an independently tunable probe laser. Ionization is not normally observed when the pump laser is tuned off of the m-photon resonance. In most cases, the high resolution of the pump laser enables the preparation of a single vibrational and rotational level of the excited state of interest; thus, it is possible to study photoionization dynamics of electronically excited states at the rovibronic state-selected level. By suitable choice of the m-photon pump transition, it is also possible to selectively ionize a species of interest without ionizing any other species that might be present. In this manner, photoelectron spectra of van der Waals molecules and free radicals can be obtained without interference from more abundant species. Another advantage of REMPI-PES is that by judicious choice of the resonant intermediate level, specific electronic states of the ion, and even specific vibrational modes, can be selectively enhanced. In summary, the characteristics of pulsed laser sources — i.e., high peak power, short pulse duration, and high wavelength resolution — provide the opportunity for the development of previously unexplored areas of photoelectron spectroscopy.

In this Chapter we will describe the major classes of molecular problems that currently are being studied using REMPI-PES. These will be illustrated, for the most part, with examples from our own work; however, an extensive table summarizing work published before October 1987 is included in the Appendix.

The outline of this Chapter is as follows. First, we will discuss the experimental techniques and instrumentation used to perform REMPI-PES studies. Second, we will focus on REMPI-PES studies of excited state photoionization dynamics. Third, we will discuss the application of REMPI-PES to the study of transient species. As discussed above, the selective ionization possible using REMPI allows the determination of photoelectron spectra for transient species such as van der Waals molecules and free radicals. Fourth, we will discuss REMPI-PES studies of larger molecules. By recording spectra for a number of different vibrational modes in the resonant intermediate level, it is possible to determine the vibrational frequencies for a large number of the normal modes of large polyatomic molecules. In addition, REMPI-PES can provide information on the radiationless decay of the intermediate state. Finally, we will conclude the Chapter with a discussion of future prospects for REMPI-PES studies.

2. EXPERIMENTAL CONSIDERATIONS

The basic apparatus for REMPI-PES studies consists of a tunable pulsed laser and a photoelectron spectrometer. In addition, a mass spectrometer is frequently used to determine the excitation spectrum of the multiphoton transition from the ground state to the resonant intermediate level. In most of the studies discussed here, the laser pulse width is 5-10 nsec long; however, work has begun using picosecond lasers.[18]

The high peak power of the pulsed laser can produce an extremely high ionization rate during the pulse, which can result in space charge broadening of the photoelectron peak.[19] Therefore, it is necessary to reduce the power of the laser until the number of electrons produced per laser shot is of the order of 100.[20] For low repetition rate lasers (10-100 shots per second), the time averaged ionization rate for REMPI-PES will be significantly lower than that normally achieved for single photon PES studies. Although gating of the detection electronics provides a substantial improvement in the signal to noise ratio, it is clearly desirable to employ an electron energy analyzer with a high collection efficiency.

Both time-of-flight and electrostatic electron energy analyzers are commonly used in REMPI-PES studies. Typical collection efficiencies for these analyzers are only ~0.05% owing to the small solid angle subtended by the entrance slit. These two types of analyzers offer complementary advantages. Using a transient digitizer with a time-of-flight analyzer, it is possible to record the entire photoelectron spectrum for each laser shot; however, the resolution is not constant with energy and degrades rapidly with increasing electron kinetic energy. The electrostatic analyzer offers constant resolution as a function of electron kinetic energy; however, the photoelectron spectrum typically must be recorded one energy at a time. In principle, data collection times can be reduced by employing a multichannel detector; however, this has not yet been done for an apparatus using pulsed laser ionization sources. Both electrostatic electron energy analyzers and time-of-flight electron energy analyzers can be very high resolution devices, particularly when used with a supersonic molecular beam source that rotationally cools the target gas and reduces the transverse velocity component of the molecular beam. Long et al.[21] recently demonstrated an electron energy resolution of 3 meV using a time-of-flight electron spectrometer with a pulsed supersonic molecular beam source. This is the highest resolution achieved to date using a dispersive analyzer.

In general, REMPI-PES studies are performed using linearly polarized laser radiation with the entrance slit of the analyzer positioned along the polarization axis of the light (i.e. at $\theta=0°$). Photoelectron angular distributions are obtained by rotating the polarization of the light while leaving the analyzer stationary. Photoelectron angular distributions following REMPI can be significantly more complicated than angular distributions following single photon ionization. Since there is no "magic angle" (the angle at which the photoelectron band intensities are proportional to the partial cross sections) for REMPI-PES,[22] the determination of partial cross sections in REMPI-PES requires the determination of the angular distribution for each band in the spectrum.

A variation of the standard time-of-flight electron energy analyzer is the magnetic bottle electron spectrometer designed by Kruit and Read.[23] This spectrometer incorporates a magnetic lens in the ionization region that parallelizes electrons with different ejection angles. As a result of this magnetic focussing, the transit time to the detector is independent of the transverse velocity component to an accuracy of better than 1% for a 10 eV electron. Hence, the instrument combines extremely high collection efficiency (50%) with the high resolution of a time-of-flight instrument, making it an ideal tool for REMPI-PES studies. The use of a transient digitizer to record the spectra further reduces the data collection time. Another aspect of this spectrometer is that the acceleration/deceleration of electrons in the weak magnetic field of the flight tube introduces no lens effect, such as that occurring in purely electrostatic lenses. Accordingly, electrostatic retardation can be used to lengthen flight times, thereby improving energy resolution without altering the 50% transmission of the device. An electron energy resolution as high as 8 meV has been obtained with an analyzer of this design,[24] and typical energy resolutions are in the range of 15-25 meV. Since this apparatus collects 50% of the electrons in the ionization region, the dynamical information contained in the photoelectron angular distributions cannot be obtained in a straightforward manner. However, the relative partial cross sections are determined directly. A more serious drawback is that the ionization occurs in a magnetic field of 1 Tesla, and under certain conditions, this high magnetic field is expected to influence the observed photoelectron spectra. It is therefore desirable to have both a magnetic bottle electron spectrometer and a field-free electron spectrometer.

A photoelectron spectrometer of extremely high resolution (0.15 meV) has been employed by Müller-Dethlefs et al.[25-27] This spectrometer is based on the detection of photoelectrons of nearly zero kinetic energy, which are produced only if the total energy of the ionizing photon(s) is equal to the energy difference between an ionic state and the initial neutral molecular state. The apparatus uses a pulsed electric field discrimination technique in combination with a steradiancy analyzer. Spectra are recorded by scanning the ionizing laser wavelength while monitoring the production of zero kinetic energy electrons. The combination of a high resolution laser with the high discrimination used in the detection of slow electrons makes it possible to study the rotational structure of a large number of molecular ions.[26,27] In contrast, the determination of rotationally resolved REMPI-PES spectra using dispersive analyzers has been limited to molecules having large rotational spacings (e.g., H_2)[28-30] or to molecular ions in high rotational levels.[31,32] A drawback of the threshold photoelectron spectrometer is that it is not possible to use it to study photoelectron branching ratios, since it only detects zero kinetic energy electrons. Thus, dispersive analyzers and threshold analyzers are destined to play complementary roles in REMPI-PES studies, just as they have in single photon UV-PES studies for the past ten years.

3. PHOTOIONIZATION DYNAMICS OF EXCITED MOLECULAR STATES

A major area of investigation using REMPI-PES is the study of branching ratios following photoionization of excited molecular states. Such studies provide considerable new information, since conventional UV-PES studies have been almost completely limited to ground state photoionization. Four examples will be discussed here, and owing to its importance as a theoretically tractable system, two of the four examples involve molecular hydrogen. In the first example, the (3+1) REMPI-PES spectra of H_2 via the C $^1\Pi_u$ state will be discussed. This example provides a particularly good illustration of the interplay between experiment and theory. In the second example, the (3+1) REMPI-PES spectra of N_2 via the b $^1\Pi_u$, c $^1\Pi_u$, and o $^1\Pi_u$ states will be discussed. This example illustrates several ways in which the character of the resonant intermediate level affects the observed REMPI-PES spectrum. In the third example, recent work on the autoionization of high Rydberg states of H_2 converging to the H_2^+ X $^2\Sigma_g^+$, $v^+=2$ ionization limit will be described. These high Rydberg states are excited by pumping two photon transitions to the E,F $^1\Sigma_g^+$, $v'=$E2 level with a first laser, and subsequently probing transitions from the E,F $^1\Sigma_g^+$, $v'=$E2 level to high Rydberg states with a second laser. The photoelectron branching ratios for autoionization are determined using a magnetic bottle electron spectrometer. This

example serves to illustrate the detailed information that is now
accessible using REMPI-PES. In the fourth example, studies of the
vibrational autoionization of NO will be discussed. In concluding
this Section, the newly developed technique of circular dichroism of
angular distributions (CDAD)[33] using REMPI-PES will be described.
This technique is being used to study atomic and molecular alignment.

3.1. Photoionization of H_2 C $^1\Pi_u$

Several years ago, we reported photoelectron spectra obtained
following (3+1) ionization via the C $^1\Pi_u$, v'=0-4 levels of molecular
hydrogen.[34] The relevant potential energy curves for the excited and
ionic states of H_2 are shown in Figure 1. The REMPI spectra for the

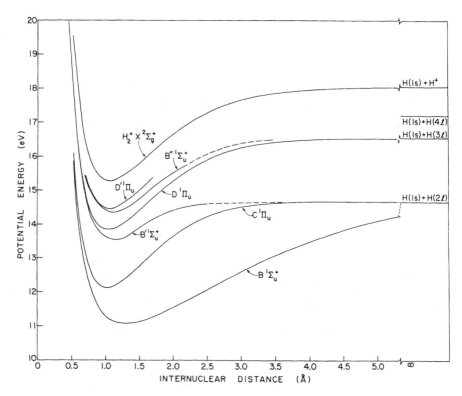

Figure 1. Relevant potential energy curves for excited states of H_2
and for H_2^+ X $^2\Sigma_g^+$.

C $^1\Pi_u$, v'=0-4 ← X $^1\Sigma_g^+$, v"=0 transitions obtained by monitoring the H_2^+ ion signal are shown in Figure 2. The C $^1\Pi_u$ Rydberg state has a potential curve similar to that of the X $^2\Sigma_g^+$ ground state of H_2^+.[35] Thus, the vibrational overlap integrals between C $^1\Pi_u$, v' and H_2^+ X $^2\Sigma_g^+$, v^+ will be nearly unity for v^+=v' and nearly zero for $v^+ \neq v'$. On the basis of the Franck-Condon principle, the photoelectron spectra are therefore expected to show strong v^+=v' peaks, with very little intensity in the $v^+ \neq v'$ peaks. Qualitatively, these expectations are fulfilled, as is seen in Figure 3, which shows the photoelectron spectra obtained following (3+1) ionization via the C $^1\Pi_u$, v'=0-4 Q(1) transitions. These spectra were obtained along the laser polarization axis (θ=0°) using an electrostatic energy analyzer. The Q(1) transitions were chosen because they access the Π_u^- component of the C $^1\Pi_u$ state, which is unperturbed by the B $^1\Sigma_u^+$ state. In each spectrum the v^+=v' peak is the most intense and the $v^+ \neq v'$ peaks are significantly weaker, in accord with the Franck-Condon arguments. However, a quantitative comparison of the relative intensities with the theoretical Franck-Condon factors of Table 1 reveals significant discrepancies, particularly for v'=3 and 4. For example, in the v'=4 spectrum the observed v^+=3, 5, and 6 peaks are too large by factors of 3, 2, and 23, respectively.[34]

Table 1. Franck-Condon factors for H_2^+ X $^2\Sigma_g^+$, v^+ ← H_2 C $^1\Pi_u$, v'

v^+	v'=0	v'=1	v'=2	v'=3	v'=4
0	9.89×10^{-1}	1.10×10^{-2}	4.67×10^{-6}	1.45×10^{-8}	2.26×10^{-9}
1	1.06×10^{-2}	9.66×10^{-1}	2.29×10^{-2}	2.04×10^{-5}	2.98×10^{-8}
2	3.08×10^{-4}	2.16×10^{-2}	9.43×10^{-1}	3.55×10^{-2}	6.47×10^{-5}
3	1.46×10^{-5}	9.98×10^{-4}	3.22×10^{-2}	9.19×10^{-1}	4.80×10^{-2}
4	1.04×10^{-6}	6.53×10^{-5}	2.13×10^{-3}	4.16×10^{-2}	8.96×10^{-1}
5	1.00×10^{-7}	5.95×10^{-6}	1.82×10^{-4}	3.71×10^{-3}	4.92×10^{-2}
6	1.22×10^{-8}	7.24×10^{-7}	2.03×10^{-5}	3.98×10^{-4}	5.69×10^{-3}
7	1.87×10^{-9}	1.12×10^{-7}	2.92×10^{-6}	5.42×10^{-5}	7.43×10^{-4}
8	3.77×10^{-10}	2.04×10^{-8}	5.34×10^{-7}	8.99×10^{-6}	1.21×10^{-5}
9	1.04×10^{-10}	4.38×10^{-9}	1.19×10^{-7}	1.82×10^{-6}	2.43×10^{-5}
10	3.75×10^{-11}	1.12×10^{-9}	3.02×10^{-8}	4.65×10^{-7}	5.68×10^{-6}

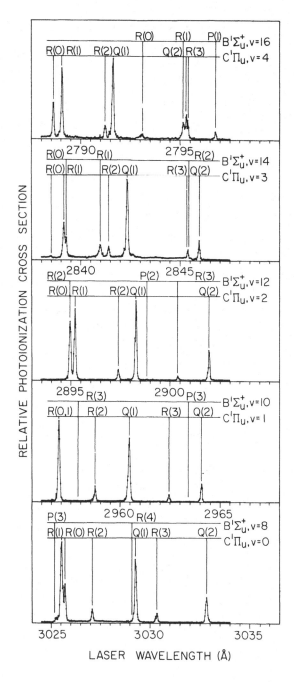

Figure 2. Relative multiphoton ionization cross sections for the production of H_2^+ from H_2 via a resonant three-photon transition to the $C\ ^1\Pi_u$, v=0-4 levels. Assignments are from the single-photon absorption data of Ref. 36.

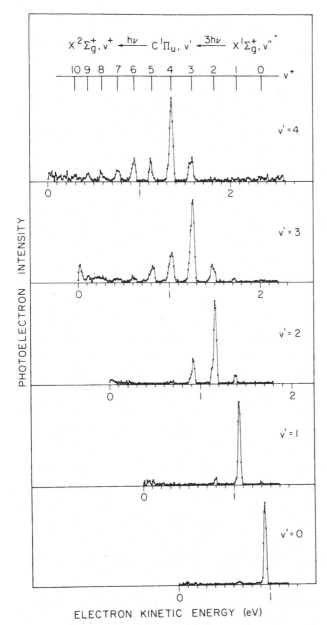

Figure 3. Photo-electron spectra of H_2 following (3+1) ionization via the C $^1\Pi_u$ v'=0-4 levels.

PHOTOELECTRON INTENSITY

ELECTRON KINETIC ENERGY (eV)

The discrepancy between the experimental results and theoretical Franck-Condon factors could have a number of sources. These include: (1) a kinetic energy dependence of the electronic transition matrix element, which must be taken into account even within the Franck-Condon approximation; (2) an internuclear distance (R) dependence of the same electronic matrix element which, by definition, constitutes a breakdown of the Franck-Condon approximation; and (3) a v^+-dependence of the photoelectron angular distributions. Dixit et al. have recently begun a theoretical program to analyze REMPI processes in diatomic molecules.[37-42] In calculations of (3+1) ionization of H_2 via the C $^1\Pi_u$ state,[37] they have included the three effects discussed above. However, while the inclusion of these effects improves the agreement with experiment, it is still not good. Recently, we have measured angle-integrated branching ratios using two different techniques.[29,30] In the first experiment,[29] the photoelectron angular distributions were determined for each spectrum and then integrated to give branching ratios. In the second experiment,[30] the integrated branching ratios were determined directly using a magnetic bottle electron spectrometer. The two measurements are in good agreement. Figure 4 shows a comparison of the angle-integrated branching ratios calculated by Dixit et al.[37] with the results of O'Halloran et al.[30] The discrepancies are quite apparent, particularly for v'=3 and 4, where the observed distribution is much broader than the theoretical predictions. This indicates that the photoionization of the C $^1\Pi_u$ state is more complicated than the direct excitation of the Rydberg electron into the X $^2\Sigma_g^+$ continuum, and suggests that another mechanism is important for higher v'. It is worth noting that the photoelectron angular distributions for the $v^+\neq v'$ photoelectron bands are generally more isotropic than those for the $v^+=v'$ bands,[29] which also suggests that another mechanism is responsible for the observed intensity of these bands.

Additional evidence for the complexity of the photoionization process is found in the rotational structure of the vibrational bands in the C $^1\Pi_u$, v'=0-4 spectra.[29,30] Before discussing this structure

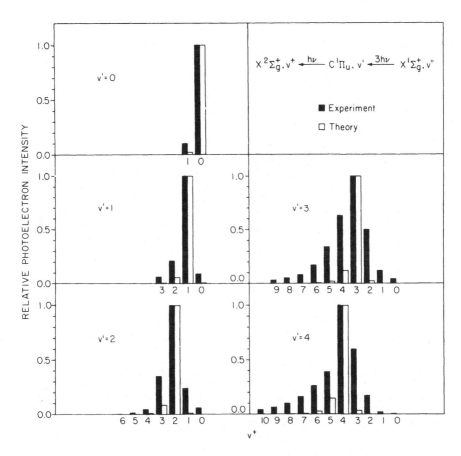

Figure 4. Vibrational branching ratios determined for (3+1) ionization of H_2 via C $^1\Pi_u$, v', Q(1) transitions. The calculation is that
of Dixit, Lynch, and McKoy (Ref. 37).

in detail, it is first appropriate to review the selection rules for
the ionizing transition. The single photon ionization step from the
C $^1\Pi_u$ state is governed by the selection rules[43] a↔/↔s, +←→−, and
$\Delta J=0,\pm 1$. Using these selection rules, the allowed final rotational
levels of H_2^+ can be readily determined using a schematic diagram
analogous to those of Herzberg.[43] Figure 5 shows the important

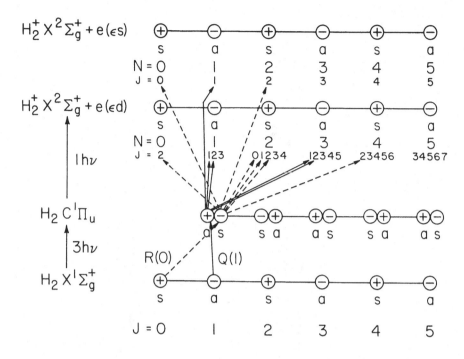

Figure 5. Schematic diagram of the allowed rotational transitions for (3+1) ionization of H_2 via the C $^1\Pi_u$ state showing the R(0) excitation (dashed lines) and the Q(1) excitation (solid lines) from the ground electronic state. The ionization continua are described in Hund's case (d).

symmetry labels and quantum numbers for the H_2 X $^1\Sigma_g^+$, H_2^* C $^1\Pi_u$, and H_2^+ X $^2\Sigma_g^+$ + ϵs and ϵd electronic states that are involved in the (3+1) REMPI process described above. The only difference between this diagram and the standard diagrams given by Herzberg[43] is that, in the present case, the ionization continuum is represented in Hund's case (d), i.e., the electrons are completely uncoupled from the molecular axis. This is the coupling case in which the observable quantity, the rotational level of H_2^+, is a good quantum number. On the basis of parity considerations, the outgoing partial wave must be even; however, only s and d waves have been included in Figure 5, as higher partial waves will almost certainly be unimportant owing to large centrifugal barriers. It is clear from the selection rules that, for single photon ionization from the C $^1\Pi_u$ state, the R(0) and Q(1) photoelectron spectra will show different sets of rotational levels.[28,29,41] Only even values of N^+ result from photoionization via R(0), while only odd values of N^+ result from photoionization via Q(1). It must be remembered that this diagram only determines which rotational levels are allowed on the basis of symmetry selection rules; it gives no information on the relative intensity of the various allowed ionizing transitions. Similar selection rules can be derived for any homonuclear molecule. In addition, analogous arguments also apply to the photoionization of heteronuclear molecules to the extent that the ejected electron can be described by all even or by all odd partial waves. For example, these arguments have aided in the interpretation of rotationally resolved REMPI-PES spectra of NO.[32]

The selection rules of Figure 5 show that for ionization via the Q(1) transition through the C $^1\Pi_u$ state, the H_2^+ ion can only be formed in the rotational levels $N^+=1$ or 3. This is confirmed by the observation of only $N^+=1$ and 3 photoelectron peaks in the C $^1\Pi_u$, $v'=0-4$, Q(1) spectra shown in Figure 6. These spectra were obtained at somewhat higher resolution than those of Figure 3 by using the magnetic bottle electron spectrometer.[30] (Note that the spectra shown in Figure 6 are angle-integrated spectra.) In these spectra, the relative intensity of the $N^+=3$ photoelectron peak increases dramatically with increasing v', both for $v^+=v'$ and for $v^+\neq v'$. In addition, for a given intermediate level, v', the $N^+=3$ rotational peaks tend to be larger relative to the $N^+=1$ peaks in the $v^+\neq v'$ bands. This definitely indicates that the photoionization mechanism is changing with increasing vibrational quantum number, v', and that this mechanism contributes to the intensity of the $v^+\neq v'$ bands.

It is sometimes useful to describe the photoionization dynamics in terms of the angular momentum transfer, j_t, which is defined as the angular momentum exchange between the unobserved initial and final

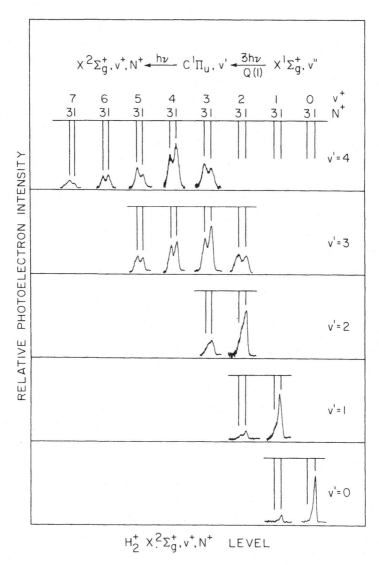

Figure 6. Photoelectron spectra determined following (3+1) ionization via the H_2 C $^1\Pi_u$, v'=0-4 ← X $^2\Sigma_g^+$, v"=0 Q(1) transitions. The spectra of individual ionic vibrational bands were recorded separately, with retarding voltages chosen so as to achieve comparable energy resolution for each vibrational band. Note that the horizontal scale does not indicate energy, although the individual vibrational bands are plotted with the same energy scale. The integrated areas of the vibrational bands are set equal to the ionic vibrational branching ratios.

angular momenta.[44-46] If only s and d partial waves are considered, the $N^+=1$ peaks in Figure 6 can only arise from $j_t=1$ processes, while the $N^+=3$ peaks can only arise from $j_t=3$ processes.[30] If the photoionization process is divided into two parts, corresponding to the initial excitation followed by the photoelectron escape, then only $j_t=1$ processes can occur in the excitation step.[45] The higher value of angular momentum transfer, $j_t=3$, must result from anisotropic interactions between the ion core and the escaping photoelectron. The data shown in Figure 6 indicate that these anisotropic interactions, and thus the $j_t=3$ processes, become increasingly important with increasing vibrational quantum number, and are relatively more important for $v^+\neq v'$.

The increasing intensity of $j_t=3$ processes with increasing v' indicates that photoionization from these levels does not proceed by the direct ejection of the Rydberg electron. If, instead, the photoionization process involves excitation to an autoionizing level at the four photon energy followed by decay into the ionization continuum, a mechanism would be provided for both the non-Franck-Condon behavior and for the increasing importance of anisotropic electron-ion interactions. On the basis of simple wavelength dependent studies performed by pumping different rotational levels within the C $^1\Pi_u$, v' bands, we have determined that it is unlikely that sharp autoionizing resonances are responsible for the observed behavior. In addition, the $\Delta v=-1$ propensity rule for vibrational autoionization[47] and the energetics for rotational autoionization suggest that neither of these processes contributes to the present observations.

Chupka[48] has recently suggested that doubly-excited states at the four photon energy will play an important role in the (3+1) ionization via the C $^1\Pi_u$ state. In particular, he has argued that the $2p\sigma_u 2p\pi_u$ doubly-excited state will have significant oscillator strength from the $1s\sigma_g 2p\pi_u$ C $^1\Pi_u$ state, and that autoionization of this doubly-excited state will lead to non-Franck-Condon vibrational branching ratios. Cornaggia et al.[49] have also suggested that doubly-excited states will be more important for multiphoton ionization via the C $^1\Pi_u$ state into the gerade continuum than for ionization via the E,F $^1\Sigma_g^+$ state into the ungerade continuum, for which they performed calculations. Independently, Hickman[50] has performed model calculations of the vibrational branching ratios following autoionization of the $2p\sigma_u 2p\pi_u$ doubly-excited state accessed by (3+1) excitation via the C $^1\Pi_u$ state. Using the $2p\sigma_u 2p\pi_u$ potential curve of Guberman,[51] Hickman[50] has obtained very encouraging agreement with the experimental results.

In a classical or semi-classical time dependent framework,[48] such a process could be viewed as the production of a wave packet near the inner turning point of the doubly excited state. As it evolves in time, the wave packet will produce an outgoing and incoming component; the latter will be reflected and subsequently interfere with the originally outgoing component. Because the doubly excited state has a finite width for autoionization, as the wave packet evolves it will have some probability for transitions into the $^2\Sigma_g^+$ continuum. In this model, the production of $v^+ < v'$ photoelectron bands arises from autoionization as the wave packet evolves to dissociaton products, while those with $v^+ < v'$ arise from autoionization of the incoming component propagating to smaller R.

This model also introduces the possibility that some of the molecules in the doubly-excited state will not autoionize, but rather will dissociate into a ground state atom and an excited state atom.[30,52-54] Excited state hydrogen atoms with n=3-5 have been observed, and may result from curve crossings of the repulsive $2p\sigma_u 2p\pi_u$ state with singly excited Rydberg states at large internuclear distance. In general, the dissociation processes for the C $^1\Pi_u^-$ levels are much weaker than the ionization processes. However, the (3+1) spectra via high-lying vibrational levels of the B $^1\Sigma_u^+$ state in the same energy region exhibit nearly complete dissociation, and the C $^1\Pi_u^+$ levels, which interact with the B $^1\Sigma_u^+$ state, generally display significantly more dissociation than the corresponding C $^1\Pi_u^-$ levels. This is clearly illustrated in Figure 7, which shows spectra in the region of the C $^1\Pi_u$, v'=2 $\leftarrow$ X $^1\Sigma_g^+$, v"=0 and B $^1\Sigma_u^+$, v'=12 $\leftarrow$ X $^1\Sigma_g^+$, v"=0 bands obtained by monitoring photoelectrons corresponding to photoionization of H_2 C $^1\Pi_u$ or B $^1\Sigma_u^+$ (upper frame) and to photoionization of H(3ℓ) produced by photodissociation of these levels (lower frame). The increased dissociation for the B $^1\Sigma_u^+$ levels may arise from two sources. First the B $^1\Sigma_u^+$ state samples a much larger range of internuclear distance than the C $^1\Pi_u$ state, and may have a significant direct photodissociation cross section. Second the $(2p\sigma_u)^2$ doubly excited state is allowed from the $1s\sigma_g2p\sigma_u$ B $^1\Sigma_u^+$ state, which could produce more dissociation than the $2p\sigma_u2p\pi_u$ state.

The relative positions of the C $^1\Pi_u$ and $2p\sigma_u2p\pi_u$ potential curves indicate that the transition will occur from the outer turning point of the C $^1\Pi_u$ state to the inner turning point of the $2p\sigma_u2p\pi_u$ state. Increasing the vibrational quantum number has two effects – the four photon energy is increased, and the outer turning point of the lower state is moved to larger R. For v'=0, both the total energy and the outer turning point are too small for transitions to the doubly-excited state to be very important. As v' is increased, both the overlap with the doubly-excited state and the energy requirement improve, and the effects attributable to the doubly-excited state become more noticeable, as is the case for v'=3 and 4. Eventually, as v' is increased further, the energy will be too high and the overlap again will be too poor for the $2p\sigma_u2p\pi_u$ state to play a role. In the region just above v'=4 the qualitative model described above corresponds to excitation to somewhat larger R than the classical inner turning point, with considerable amplitude for autoionization at smaller internuclear distances. This would lead to an increase in the population of vibrational levels of H_2^+ with $v^+<v'$. As is seen by the spectra shown in Figure 8, these arguments are supported by the photoelectron spectra obtained at θ=0° following (3+1) ionization via the C $^1\Pi_u$, v'=5,6 Q(1) transitions. In both spectra, the $v^+=v'$ peak is the largest, with much smaller $v^+\neq v'$ peaks.

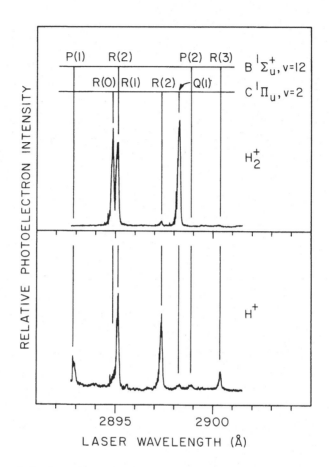

Figure 7. Relative photoelectron intensity recorded as a function of laser wavelength. The spectra were measured with a detector gated on the time of arrival of either electrons resulting from photoionization of H_2 C $^1\Pi_u$ (upper trace) or electrons resulting from photoionization of H(31) (lower trace). The vertical scales of the two traces are arbitrary. Assignments of three-photon resonances with C $^1\Pi_u$, $v'=2 \leftarrow$ X $^1\Sigma_g^+$, $v''=0$ or B $^1\Sigma_u^+$, $v'=12 \leftarrow$ X $^1\Sigma_g^+$, $v''=0$ are from the data of Ref. 55.

Comparison with Figure 3 reveals the distribution of $v^+ \neq v'$ peaks shifts to smaller values of v^+ with increasing v'. In addition, the sum of the intensities of the $v^+ \neq v'$ peaks relative to the $v^+=v'$ peak decreases monotonically as v' is increased from 4 to 6. The C $^1\Pi_u$,

$v'=7,8 \leftarrow X\ ^1\Sigma_g^+$, $v'=0$ bands are overlapped by the much more intense $B'\ ^1\Sigma_u^+$, $v'=1,2 \leftarrow X\ ^1\Sigma_g^+$, $v'=0$ bands, and could not be studied. Although the $C\ ^1\Pi_u$, $v'=9$ Q(1) transition is blended with the $D\ ^1\Pi_u$, $v'=1$ Q(1) transition, some information can nevertheless be obtained regarding ionization of the $C\ ^1\Pi_u$, $v'=9$ level. The photoelectron spectrum following (3+1) ionization at this wavelength[56] is shown in

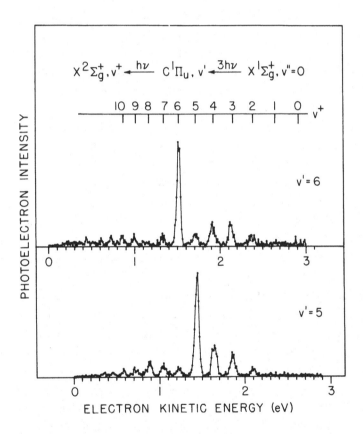

Figure 8. Photoelectron spectra of H_2 determined along the laser polarization axis ($\theta=0°$) at the wavelengths of the resonant three-photon $C\ ^1\Pi_u$, $v'=5, 6 \leftarrow X\ ^1\Sigma_g^+$, $v''=0$, Q(1) transitions.

the center frame of Figure 9. The D $^1\Pi_u$ state corresponds to the
$1s\sigma_g3p\pi_u$ Rydberg state, and the v'=1 photoelectron spectrum is similar
to that of the C $^1\Pi_u$, v'=1 level. However, the small v'=9 (at ～ 1.7
eV) peak almost certainly corresponds to ionization via the C $^1\Pi_u$,
v'=9 level. Although the v'=9 peak is weak, it is interesting to note
that no intensity is observed for v^+=5-12, which suggests that at this
energy and internuclear distance, the $2p\sigma_u2p\pi_u$ state no longer plays
an important role in the ionization process.

Doubly-excited states at the four photon energy are also expected
to play a role in the (3+1) ionization via the C $^1\Pi_u$ states of the
heavier isotopes of H_2. In particular, we have recently recorded the
photoelectron spectra following (3+1) ionization via the C $^1\Pi_u$, v'=0-4
levels of D_2.[57] The REMPI spectra obtained by monitoring the D_2^+ ion
signal are shown in Figures 10 and 11. The REMPI-PES spectra for the
C $^1\Pi_u$, v'=0-3 ← X $^1\Sigma_g^+$, v"=0, Q(3) transitions are shown in Figure 12,
and the REMPI-PES spectrum for the C $^1\Pi_u$, v'=4 ← X $^1\Sigma_g^+$, v"=0, Q(1)
transition is shown in Figure 13. As in H_2, the v^+=v' peak dominates
each spectrum. The most striking difference between the H_2 and D_2
C $^1\Pi_u$ photoelectron spectra is that for v'=3 and 4, the H_2 spectra
extend to v^+=v'+6, while the D_2 spectra show significant intensity
only for v^+≤v'+2. Although there are several possible explanations
for this observation, they are all consistent with the model involving
the $2p\sigma_u2p\pi_u$ doubly-excited state. The doubly-excited potential
curves for H_2 and D_2 will be nearly identical, and the mass effect on
the electronic autoionization width should be small. Since the four
photon energies for excitation to v'=3 and 4 are smaller in D_2 than in
H_2 (by ～0.26 and 0.34 eV, respectively), excitation to the doubly-
excited curve energetically is less favorable in D_2 than in H_2. In
addition, for lower values of v', the smaller range of R sampled by
the D_2 vibrational wavefunctions in the C $^1\Pi_u$ state will decrease the
vibrational overlap with the $2p\sigma_u2p\pi_u$ state. Finally, if the D_2 is
excited to the same position on the doubly-excited potential curve as
the H_2, the D_2 wave packet will propagate at only $1/\sqrt{2}$ of the speed of
the H_2 wave-packet. Thus, with the same autoionization rate, a much

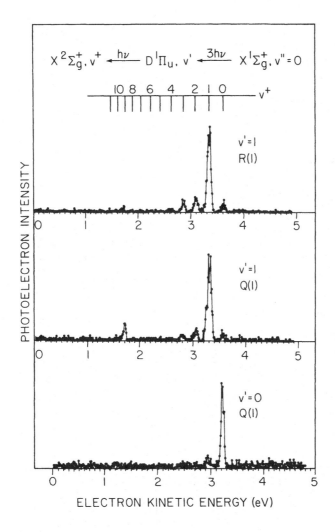

Figure 9. REMPI-PES spectra of H_2 determined at the wavelengths of the resonant three photon $D\ ^1\Pi_u$, $v'=0 \leftarrow X\ ^1\Sigma_g^+$, $v''=0$, Q(1); $D\ ^1\Pi_u$, $v'=1 \leftarrow X\ ^1\Sigma_g^+$, $v''=0$, Q(1); and $D\ ^1\Pi_u$, $v'=1 \leftarrow X\ ^1\Sigma_g^+$, $v''=0$ R(1) transitions.

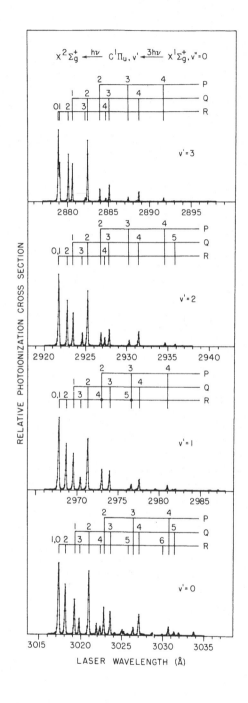

Figure 10. The (3+1) ionization spectra of D_2 via the C $^1\Pi_u$, v'=0-3 levels obtained by monitoring the D_2^+ ion signal. Assignments are from the single photon absorption data of Ref. 58.

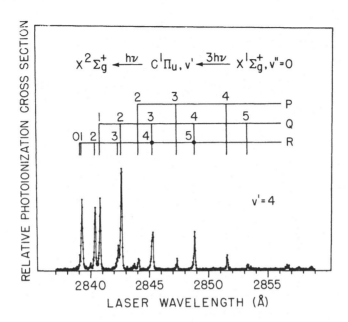

Figure 11. The (3+1) ionization spectrum of D_2 via the C $^1\Pi_u$, v'=4 level obtained by monitoring the D_2^+ ion signal. Assignments are from the single photon absorption data of Ref. 58.

narrower envelope is expected for D_2. Of course, detailed calculations of the C $^1\Pi_u$ photoelectron spectra are necessary to determine the validity of these arguments for D_2. However, at least qualitatively, it appears that the D_2 spectra can be explained in a manner consistent with the H_2 spectra.

In summary, it does not appear that the existing REMPI-PES data on the (3+1) ionization of H_2 via the C $^1\Pi_u$ state can be explained in terms of the simple direct photoionization of the Rydberg electron into the H_2^+ X $^2\Sigma_g^+$ continuum. As suggested by Chupka[48] and by Hickman,[50] excitation and subsequent autoionization of the $2p\sigma_u 2p\pi_u$ doubly excited state appear to strongly influence the vibrational branching ratios, particularly for C $^1\Pi_u$, v'=3-6. Although a direct experimental study of the $2p\sigma_u 2p\pi_u$ state remains to be performed, the

94

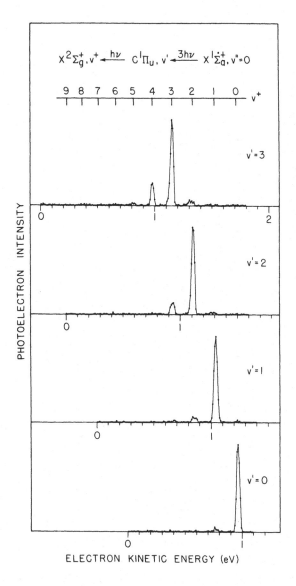

Figure 12. Photoelectron spectra of D_2 determined along the laser polarization axis at the wavelengths of the resonant three photon C $^1\Pi_u$, v'=0-3 ← X $^1\Sigma_g^+$, v"=0, Q(3) transitions.

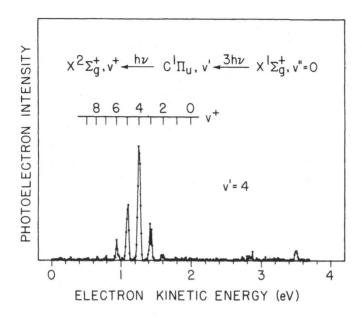

PHOTOELECTRON INTENSITY

$X^2\Sigma_g^+, v^+ \xleftarrow{h\nu} C^1\Pi_u, v' \xleftarrow{3h\nu} X^1\Sigma_g^+, v''=0$

8 6 4 2 0 v^+

$v'=4$

ELECTRON KINETIC ENERGY (eV)

Figure 13. Photoelectron spectrum of D_2 determined along the laser polarization axis at the wavelength of the resonant three photon $C^1\Pi_u$, $v'=4 \leftarrow X^1\Sigma_g^+$, $v''=0$ Q(1) transition. The peaks at 2.85 and 3.51 eV correspond to the single photon ionization of D^* (n=3) and D^* (n=4), respectively. These excited state atoms are produced by single photon dissociation of the $C^1\Pi_u$ state.

present data serve to bracket the position of this state. When coupled with more detailed calculations of the vibrational branching ratios, these data should improve the understanding of the doubly excited states of H_2.

3.2 Photoionization of Excited States of N_2

The vibrational selectivity observed for the (3+1) REMPI of H_2 via the $C^1\Pi_u$ state is a manifestation of the Franck-Condon principle, and has been seen in many other species, including NO,[59] N_2,[60] NH_3,[59,61,62] benzene,[21,63] toluene,[19] and aniline.[64] Of these, the study of N_2 is of interest since it demonstrates that it is possible to produce electroncially excited, vibrationally state-selected molecular ions by REMPI via a Rydberg state with an electronically excited ion core.[60] Specifically, it demonstrates that the dominant ionization pathway for (3+1) ionization via the o_3 $^1\Pi_u$, $v'=1,2$ levels of N_2 leads to the production of N_2^+ A $^2\Pi_u$, $v^+=1,2$, respectively.

The relevant potential curves of N_2 and N_2^+ are shown in Figure 14. The o_3 $^1\Pi_u$ state of N_2 has the electron configuration $\ldots(1\pi_u)^3(3\sigma_g)^2 3s\sigma_g$ and is the lowest member of Worley's third Rydberg series,[66] which converges to N_2^+ A $^2\Pi_{1/2u}$. It is expected that the

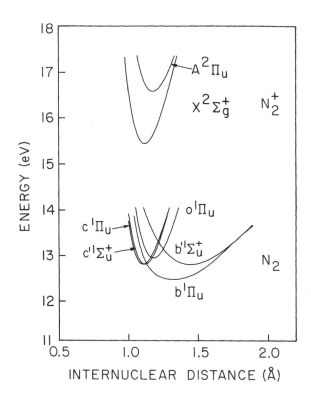

Figure 14. Relevant potential energy curves of N_2 and N_2^+. The N_2 curves were generated from the deperturbed parameters of Ref. 65.

ionizing transition from the o_3 $^1\Pi_u$ state will strongly favor the removal of the outer $3s\sigma_g$ electron, leading to the direct production of N_2^+ A $^2\Pi_u$. Similar behavior has been observed in a number of atomic systems, that is, the photoionization of Rydberg states leads to the ejection of the Rydberg electron without a change in the electronic state of the ion core.[67] In addition to preserving the electronic state of the ion core, the A $^2\Pi_u$ ← o_3 $^1\Pi_u$ ionizing transition is also expected to preserve the vibrational level of the o_3 $^1\Pi_u$ Rydberg state. Franck Condon factors calculated using Morse potentials for both states[35,65] are greater than 0.99 for the $v^+=v'$ transition for both $v'=1$ and 2.

Photoelectron spectra were recorded at the wavelengths corresponding to the three photon transitions to the R-branch bandhead of the o_3 $^1\Pi_u$, $v'= 2$ ← X $^1\Sigma_g^+$, $v''=0$ band and to three positions, including the R-branch bandhead within the o_3 $^1\Pi_u$, $v'=1$ ← X $^1\Sigma_g^+$, $v''=0$ band. Figure 15 shows the photoelectron spectra obtained via the o_3 $^1\Pi_u$, $v'=1$ and 2 bands. Of the three spectra obtained via the $v'=1$ band, that shown in Figure 15 displays the greatest fraction (~20%) of photoelectrons with $v^+\neq v'$. At other wavelengths, N_2^+ A $^2\Pi_u$, $v^+=1$ is produced with ≥90% purity. It is clear that the dominant ionization process corresponds to excitation of the Rydberg electron, with preservation of the electronic and vibrational state of the ion core. In addition, a large fraction of those ions produced by core-switching transitions undergo a change only in vibrational state, while retaining the A $^2\Pi_u$ ion core.

The ability to prepare state-selected ions offers the exciting possibility of studying ion-molecule reactions as a detailed function of the internal energy of the ionic reactant. Because REMPI allows the selective ionization of one component of a complex mixture, state preparation of ions and the study of their subsequent reactions can be performed in the same region of space. This would allow the study of reactions of ions in relatively short-lived electronic states without interference from extraneous ionization products. Although no studies

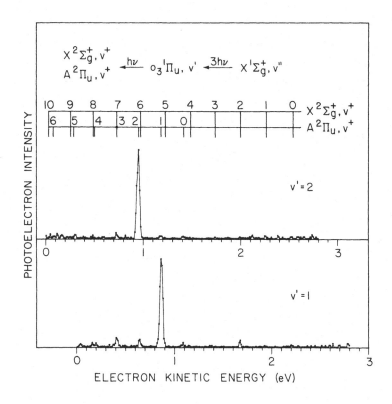

Figure 15. Photoelectron spectra obtained following (3+1) ionization of N_2 via the $o_3\,^1\Pi_u$, v'=1, 2 levels. The energy levels of N_2^+ were obtained from Ref. 66.

using electronically excited ions have been performed so far, the first studies using vibrationally state-selected NH_3^+ have recently been reported.[62,68,69]

The effects of Rydberg-valence interactions in the resonant intermediate level on REMPI-PES spectra have been the subject of a number of studies. In the first of these, White et al.[70] discussed perturbations in the C $^2\Pi$ state of NO and their influence on the REMPI-PES spectra. The b $^1\Pi_u$, c $^1\Pi_u$, and o $^1\Pi_u$ states of N_2 provide another example of the effects of Rydberg-valence interactions.[71] The potential curves for the b $^1\Pi_u$ and c $^1\Pi_u$ states are included in Figure

14. The b $^1\Pi_u$ state is a valence state with two dominant configurations,[72] corresponding to $...(2\sigma_u)^2(1\pi_u)^3(3\sigma_g)^1(1\pi_g)^2$ and $...(2\sigma_u)^1(1\pi_u)^4(3\sigma_g)^2(1\pi_g)^1$, and the c $^1\Pi_u$ state is a Rydberg state with a ground state ion core and the dominant configuration $...(2\sigma_u)^2(1\pi_u)^4(3\sigma_g)^1 3p\pi_u$.[66] In the spectral region of interest, however, the b $^1\Pi_u$ and c $^1\Pi_u$ states are strongly mixed by a homogeneous perturbation. Recently these perturbations have been extensively analyzed at the vibronic level by Stahel et al.,[65] and this analysis is quite useful in understanding the photoelectron branching ratios from several of these strongly perturbed levels.

In the analysis of Stahel et al.,[65] the eigenstates Ψ are written as a linear combination of diabatic wavefunctions

$$\sum_e \sum_v c_{ev} \, \phi_e \chi_{ev}(R). \tag{1}$$

Here the ϕ_e are the diabatic electronic wavefunctions, the $\chi_{ev}(R)$ are the vibrational wavefunctions calculated in the diabatic potentials, and the c_{ev} are the R-independent coefficients. The observed photoelectron spectrum following REMPI is then determined by the partial photoionization cross sections from the state Ψ into the various vibronic continua $\Psi_{e^+v^+} \phi_{e^-}$, where e^+v^+ labels the electronic and vibrational state of the ion produced and ϕ_{e^-} is the continuum electronic wavefunction. In the Franck-Condon approximation, the partial photoionization cross section to the e^+v^+ state is proportional to

$$|\langle\Psi|\mu|\Psi_{e^+v^+}\phi_{e^-}\rangle|^2 = |\sum_e \sum_v c_{ev}\langle\phi_e|\mu|\phi_{e^+v^+}\phi_{e^-}\rangle \langle\chi_{ev}|\chi_{e^+v^+}\rangle|^2. \tag{2}$$

Although the c_{ev} are known from the work of Stahel et al.[65] and the $\langle\chi_{ev}|\chi_{e^+v^+}\rangle$ are readily calculated, the matrix elements $\langle\phi_e|\mu|\phi_{e^+v^+}\phi_{e^-}\rangle$ present two problems. First, these matrix elements depend on the kinetic energy of the ejected electron, and therefore depend on the vibrational state of the ion produced. At the relatively low kinetic energies treated here, it is probably not reasonable to assume they are independent of v and v^+. However, in the absence of resonance phenomena it is reasonable to assume that these matrix elements will vary smoothly with v and v^+. Second, the specific form of the diabatic wavefunction ϕ_e is not uniquely defined. However, as a first approximation, it is reasonable to associate each diabatic wavefunction with a single electron configuration, as has been done above.[73] (In the case of the b $^1\Pi_u$ valence states two electron configurations are considered.[72]) The X $^2\Sigma_g^+$ and A $^2\Pi_u$ states of N_2^+ are also approximated by a single electron configuration. These approximations allow one to make some qualitative predictions about the electronic matrix elements and photoelectron branching ratios. For example, photoionization from the c $^1\Pi_u$ state is expected to favor the production of N_2^+ X $^2\Sigma_g^+$, which requires a one electron transition, over the production of N_2^+ A $^2\Sigma_u$, which requires a two electron transition. Similarly, the photoionization cross sections from the b $^1\Pi_u$ state to both the X $^2\Sigma_g^+$ and A $^2\Pi_u$ states are expected to be small, as they each require a two electron transition. How well the experimental results agree with these predictions will be discussed below.

Franck-Condon factors for the X $^2\Sigma_g^+$, $v^+ \leftarrow$ b $^1\Pi_u$, v', and X $^2\Sigma_g^+$, $v^+ \leftarrow$ c $^1\Pi_u$, v' ionizing transitions are given in Tables 2 and 3, respectively. The X $^2\Sigma_g^+$, $v^+ \leftarrow$ c $^1\Pi_u$, v' Franck-Condon factors were calculated using Morse potentials derived from the constants of Huber and Herzberg[35] for the X $^2\Sigma_g^+$ state and from the deperturbed diabatic constants of Stahel et al.[65] for the c $^1\Pi_u$ states. Because the deperturbed $\omega_e x_e$ value for the b $^1\Pi_u$ state is negative,[65] a Morse potential cannot be constructed for this state. Consequently, the X $^2\Sigma_g^+$, $v^+ \leftarrow$ b $^1\Pi_u$, v' Franck-Condon factors were calculated using a Morse potential for the X $^2\Sigma_g^+$ state (as before) and using a numerical potential fitted to the RKR potential of Stahel et al.[65] for the b $^1\Pi_u$ state. Numerical integration of this potential reproduces the $\Delta G_{v+1/2}$ values of Stahel et al.[65] to within 2 cm^{-1} except for

Table 2. Franck-Condon factors for N_2^+ X $^2\Sigma_g^+$, v^+ ← N_2 b $^1\Pi_u$, v'.

X $^2\Sigma_g^+$, v^+	$v'=0$	$v'=1$	$v'=2$	$v'=3$	$v'=4$	$v'=5$
0	0.0063	0.0300	0.0733	0.1203	0.1506	0.1562
1	0.0247	0.0820	0.1262	0.1104	0.0531	0.0078
2	0.0550	0.1178	0.0935	0.0204	0.0033	0.0445
3	0.0958	0.1117	0.0264	0.0067	0.0578	0.0619
4	0.1219	0.0703	0.0002	0.0538	0.0562	0.0060
5	0.1403	0.0234	0.0285	0.0657	0.0077	0.0178
6	0.1417	0.0004	0.0665	0.0270	0.0098	0.0530

Table 3. Franck-Condon factors for N_2^+ X $^2\Sigma_g^+$, v^+ ← N_2 c $^1\Pi_u$, v'.

X $^2\Sigma_g^+$, v^+	$v'=0$	$v'=1$
0	0.9705	0.0293
1	0.0274	0.9258
2	0.0020	0.0398
3	0.0001	0.0048
4	0.0000	0.0002
5	0.0000	0.0000
6	0.0000	0.0000
7	0.0000	0.0000

the $\Delta G_{1/2}$ value, which deviates by 9 cm^{-1}. This discrepancy is probably due to our improper interpolation at the bottom of the well between the RKR points for the v=0 level, which results in a slight lowering of the energy for the v=0 level.

The photoelectron spectra obtained at the R-branch bandheads of the b $^1\Pi_u$, v'=0-5 ← X $^1\Sigma_g^+$, v''=0 three photon transitions are shown in Figure 16 and the corresponding spectra for the c $^1\Pi_u$, v'=0, 1 ← X $^1\Sigma_g^+$, v''=0, transitions are shown in Figure 17. According to the analysis of Stahel et al.[65] the b $^1\Pi_u$, v'=0-2 vibronic levels contain

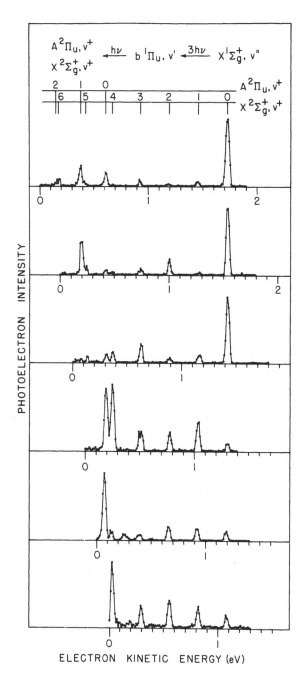

Figure 16. Photo-electron spectra of N_2 following (3+1) ionization of N_2 via the b $^1\Pi_u$, v'=0-5 levels. The impurity peaks at 0.26 and 0.50 eV in the v'=1 spectrum are due to (2+1) ionization of O_2.

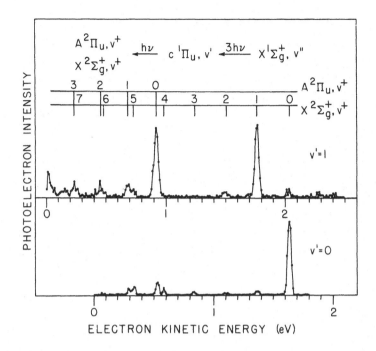

Figure 17. Photoelectron spectra of N_2 following (3+1) ionization of N_2 via the c $^1\Pi_u$, v'=0, 1 levels.

very little c $^1\Pi_u$ character, and are therefore relatively unperturbed, although the b $^1\Pi_u$, v'=2 level shows indication of predissociation in the region of the bandhead.[66,74] Thus, the photoelectron spectra obtained via all three levels should reflect relatively pure b $^1\Pi_u$ character. As is expected, photoionization from these levels popu- lates a broad distribution of vibrational levels, in qualitative agreement with the corresponding Franck-Condon factors given in Table 2. On a quantitative level, however, the agreement is rather poor. The results suggest that either a minor component of the b $^1\Pi_u$ wavefunction is responsible for the observed behavior, or that a two electron process (such as autoionization) is occurring. The latter could easily result in a breakdown of the Franck-Condon approximation. Other phenomena, such as dependence of the electronic photoionization matrix element on electron kinetic energy and internuclear distance, and a v^+ dependence of the photoelectron angular distributions might also contribute to the observed behavior.[34,75,76] The importance of such phenomena is difficult to assess for the present results due to the complex nature of the intermediate state.

Beginning with the b $^1\Pi_u$, v=1 photoelectron spectrum, the A $^2\Pi_u$, v^+=0 state of N_2^+ becomes energetically accessible and is observed with a large intensity, as it is in the b $^1\Pi_u$, v=2 photoelectron spectrum. Unfortunately, relative intensities at such low electron kinetic energies are not reliable, making it difficult to accurately compare the intensities of the A $^2\Pi_u$, v^+=0 ← b $^1\Pi_u$, v'=1 and X $^2\Sigma_g^+$, v^+=0 ← b $^1\Pi_u$, v'=1 ionizing transitions.

Unlike the b $^1\Pi_u$, v=0-2 photoelectron spectra, which exhibit a number of intense peaks, the b $^1\Pi_u$, v'=3-5 photoelectron spectra each display a single intense v^+=0 peak, with very little intensity in any of the other peaks. This is somewhat surprising since the Franck-Condon factors predict a number of other moderately intense peaks. The single intense v^+=0 peak suggests that photoionization from these levels may be dominated by a v'=0 Rydberg state component in the wavefunction of the intermediate state,[60] since photoionization from a Rydberg level is expected to preserve the vibrational quantum number of the intermediate state. The analysis of Stahel et al.[65] shows that the b $^1\Pi_u$, v'=3-5 wavefunctions contain 5.8%, 16.0%, and 22.1% c $^1\Pi_u$, v'=0 character, respectively, and this admixture could account for the intense v^+=0 peak observed in the b $^1\Pi_u$, v'=3-5 photoelectron spectra. Although this qualitative argument ignores the interference terms in Equation (2), which could lead to dramatically different behavior, the argument is a reasonable first approximation if the electronic photoionization matrix element from the c $^1\Pi_u$ state is much larger than that from the b $^1\Pi_u$ state. The latter is not unreasonable, as photoionization from the c $^1\Pi_u$ state requires only an allowed single electron transition, while photoionization from the b $^1\Pi_u$ state requires a nominally forbidden two electron transition.[72] The weaker structure in the b $^1\Pi_u$, v'=3-5 and c $^1\Pi_u$, v'=0 photoelectron spectra varies with the intermediate state. A quantitative analysis of this structure would require determining the electronic photoionization matrix elements and taking into account the interference effects.

As was mentioned above, a number of the b $^1\Pi_u$, v' levels are predissociated, the b $^1\Pi_u$, v'=3 levels being the most strongly affected.[66] This predissociation might be manifested in the photoelectron spectra as an increased intensity of a broad distribution of states, due to photoionization during the predissociation process. However, the present spectra show no evidence for such an effect, which may indicate a small photoionization cross section for the state causing the predissociation. In the present case, that state is thought to be the C' $^3\Pi_u$ state,[77] which dissociates to $N(^2D^o) + N(^4S^o)$. If one ascribes the small intensity of the N_2 b $^1\Pi_u$, v'=3 REMPI spectrum to competition by predissociation, one would expect a substantial concentration of $^2D^o$ and $^4S^o$ N atoms in the laser focus. Subsequent nonresonant ionization of the N $^2D^o$ atoms requires three of the present laser photons, while ionization of the N $^4S^o$ atoms requires four. The photoelectron peak corresponding to nonresonant ionization of N $^2D^o$ is calculated to occur at 0.60 eV in the b $^1\Pi_u$, v'=3 spectrum; however, no evidence for it is observed. Further studies are being considered to attempt resonant multiphoton ionization of the $^2D^o$ and $^4S^o$ N atoms using a second tunable laser.

While the c $^1\Pi_u$, v'=0 photoelectron spectrum shown in Figure 17 can be understood by assuming that photoionization from a Rydberg level preserves the vibrational quantum number and the electronic state of the ion core, the c $^1\Pi_u$, v'=1 photoelectron spectrum cannot be viewed in this way. Although the latter spectrum does display a prominent N_2^+ X $^2\Sigma_g^+$, v^+=1 peak, the A $^2\Pi_u$, v^+=0 peak is equally intense. Production of the A $^2\Pi_u$ electronic state from the c $^1\Pi_u$ state requires a two electron transition. The appearance of the strong A $^2\Pi_u$, v^+=0 peak can be rationalized by the existence of 7% o_3 $^2\Pi_u$, v'=0 state character in the c $^1\Pi_u$, v'=1 wavefunction. Similarly, the appearance of the A $^2\Pi_u$, v^+=1 peak in the photoelectron spectrum may be due to the existence of 2% o $^2\Pi_u$, v'=1 character in the c $^1\Pi_u$, v'=1 wavefunction. However, the large relative intensity of the A $^2\Pi_u$ peaks with respect to the X $^2\Sigma_g^+$ peaks is more difficult to understand, unless the A $^2\Pi_u$ ← o_3 $^1\Pi_u$ electronic matrix element is very much larger than the X $^2\Sigma_g^+$ ← c $^1\Pi_u$ matrix element. It should be noted that if the A $^2\Pi_u$ ← c $^1\Pi_u$ and X $^2\Sigma_g^+$ ← o_3 $^1\Pi_u$ electronic transition matrix elements are very small, as is quite likely, the interference terms in Equation (2) will become unimportant.

In summary, the photoelectron spectra obtained following (3+1) ionization of N_2 via the b $^1\Pi_u$, v'=0-5 and c $^1\Pi_u$, v'=0, 1 levels exhibit complicated structure due to perturbations of the resonant intermediate level. The vibronic analysis of the $^1\Pi_u$ and $^1\Sigma_u^+$ states of N_2 by Stahel et al.[65] provides a useful framework for understanding these photoelectron spectra by providing a breakdown of the strongly perturbed intermediate vibronic levels in terms of deperturbed diabatic states with well defined Rydberg or valence character. Thus, the strong X $^2\Sigma_g^+$, v^+=0 photoelectron peak in the spectra recorded via the valence b $^1\Pi_u$, v'=3-5 levels can be understood in terms of an admixture of c $^1\Pi_u$, v'=0 Rydberg character in the b $^1\Pi_u$ wavefunction. The admixture of valence character into a Rydberg state, however, does not appear to have as dramatic an effect. This observation indicates that, in the present case, the photoionization cross section from the valence state is smaller than from the Rydberg state, at least at the wavelengths used in this work.

This conclusion is consistent with the o_3 $^1\Pi_u$ spectra discussed above. In the analysis of Stahel et al.,[65] the o_3 $^1\Pi_u$ v'=1,2 vibronic levels are found to contain 8.4% and 10.0% b $^1\Pi_u$ character, respectively, and somewhat less c $^1\Pi_u$ character. In addition, a rotational analysis by Yoshino et al.[78] of the single photon absorption data for the o_3 $^1\Pi_u \leftarrow X$ $^1\Sigma_g^+$ transition indicates much stronger mixing at the R-branch bandhead of the v'=1 band due to a perturbation by the b $^1\Pi_u$ v'=12 level. One would at first expect these perturbations to strongly affect the vibrational branching ratios following multiphoton ionization. In particular, photoionization of high vibrational levels of the b $^1\Pi_u$ state, which is a valence state, would be expected to populate a wide distribution of N_2^+ vibrational levels in both the X $^2\Sigma_g^+$ and A $^2\Pi_u$ states. It is clear from Figure 15 that this is not the case. However, because both of the configurations important in describing the b $^1\Pi_u$ state differ from the X $^2\Sigma_g^+$ and A $^2\Pi_u$ states of N_2^+ by two orbitals, the dominance of the o_3 $^1\Pi_u$ character is not unreasonable.

3.3 Two Color Studies of Autoionizing States of H_2

The combination of the technique of optical-optical double resonance (OODR) with REMPI-PES provides an extremely powerful probe of the dynamics of autoionizing levels. In a typical experiment, the pump laser is used to prepare a single rotational, vibrational level of a low-lying electronic state, and the second laser is used to probe transitions from this state to quasi-discrete levels above the ionization potential. Electron energy analysis is then used to determine the decay dynamics following autoionization of the quasi-discrete levels. The example of OODR-REMPI-PES that will be discussed in detail here involves pumping two photon transitions from the H_2 X $^1\Sigma_g^+$, v"=0 level to the E,F $^1\Sigma_g^+$, v'=E2 level and probing single photon transitions from the E,F $^1\Sigma_g^+$, v'=E2 level to high lying np Rydberg levels near the H_2^+ X $^2\Sigma_g^+$, v+=2 threshold. The autoionizing np levels are the same as those accessed from the ground state by single photon excitation, which have been observed previously in both photo-absorption[79-86] and photoionization.[87-91]

Several OODR studies of the np Rydberg states of H_2 accessed via the E,F $^1\Sigma_g^+$ state have been performed previously using ionization detection alone.[92-95] The study of these levels using REMPI-PES techniques is of particular interest, because in recent years, the progress in theoretical studies of the photoionization dynamics of H_2 has surpassed the corresponding progress in experimental studies. Stimulated by the high resolution experimental absorption spectra of Herzberg and Jungen[86] and the photoionization spectra of Dehmer and Chupka,[90] Jungen and co-workers have developed the multichannel quantum defect theory of Seaton[96] and Fano[97] to reproduce and predict rotational and vibrational autoionization,[98,99] predissociation,[100-102] partial ionization cross-sections,[103] and the wavelength dependence of photoelectron branching ratios and angular distributions.[104,105] These calculations have been remarkably successful in reproducing the lineshapes and complex resonances observed in the high resolution photoionization spectra of Dehmer and Chupka,[90] More detailed tests of the predictions of MQDT are now necessary.

3.3.1 Vibrational Autoionization of np Rydberg States of H_2

In the spectral region that we will consider first, vibrational autoionization, which occurs through the conversion of vibrational energy of the ion core to kinetic energy of the ejected electron, is the dominant decay mechanism. Berry first considered the problem of vibrational autoionization in diatomic molecules using perturbation theory within the Born-Oppenheimer framework;[47,106-109] he found that the autoionization rate is largest for $\Delta v=-1$ transitions and decreases rapidly for transitions involving greater changes of Δv. When autoionization via $\Delta v=-1$ transitions is energetically forbidden, the process with the minimum change of Δv is expected to dominate. Recent MQDT calculations by Jungen and co-workers have gone far beyond the simple propensity rule predictions.[98-105] For example, it was found that the vibrational branching ratios can vary dramatically across an autoionizing resonance.[103-105]

Very few experimental studies have even attempted to determine final state branching ratios following autoionization of the np

Rydberg states of H_2, as the resolution and intensity available in single photon studies is generally not sufficient to permit them. Berkowitz and Chupka[110] have measured the distribution of photoelectrons from several prominent autoionizing peaks using a retarding field electron energy analyzer; however, in most cases the "peak" was actually two or more unresolved levels, complicating the interpretation of the results. Later, Dehmer and Chupka[111] indirectly determined the final vibrational state distribution following decay of approximately 100 autoionizing states by measuring the relative rates of the reactions $H_2^+ + He(Ne) \rightarrow HeH^+(NeH^+) + H$, reactions that are very sensitive to the vibrational level of H_2^+. It was found that vibrational autoionization proceeds predominantly (>95%) via $\Delta v=-1$ transitions whenever energetically possible; when $\Delta v=-1$ transitions are energetically forbidden, autoionization proceeds predominantly (>75%) via the minimum change in v. While the results of these studies are in accord with the vibrational propensity rule, the resolution and sensitivity were insufficient to perform a direct study of the mechanism of vibrational autoionization with the detail necessary to rigorously test the MQDT predictions. Using the OODR excitation scheme described above and a magnetic bottle electron spectrometer, the autoionization dynamics of individual rovibronic levels can be studied with unprecedented detail.

Before discussing the two color spectra it is first useful to review the basic spectroscopy of the Rydberg states of H_2. Since the symmetry of the X $^1\Sigma_g^+$ ground state and the E,F $^1\Sigma_g^+$ state are identical, single photon transitions from the E,F $^1\Sigma_g^+$ state access the same Rydberg states as single photon transitions from the ground state. The lower members of the Rydberg series correspond to $np\sigma$ $^1\Sigma_u^+$ and $np\pi$ $^1\Pi_u$ states; for excitation from E,F $^1\Sigma_g^+$, v'=2, J'=1, there are only four allowed series of transitions corresponding to P(1)$np\sigma$ $^1\Sigma_u^+$, R(1)$np\sigma$ $^1\Sigma_u^+$, Q(1)$np\pi$ $^1\Pi_u$, and R(1)$np\pi$ $^1\Pi_u$.

As n increases there is a transition from Hund's case (b) in which the the orbital angular momentum ℓ is strongly coupled to the

internuclear axis to Hund's case (d) in which the orbital angular
momentum ℓ is strongly coupled to the axis of rotation. For large n,
the separation into $^1\Sigma_u^+$ and $^1\Pi_u$ states no longer is observed, but
rather the rotational levels of the Rydberg states are ordered solely
on the basis of the rotational level of the ion core to which they
converge.[43] Numerous perturbations occur between the two R(1) series
($\Delta v=0$ interactions) resulting in significant alterations of level
positions, line intensities, and linewidths. At high principal
quantum number ($n \gtrsim 8$), these states are no longer labeled $np\sigma$ and $np\pi$,
but rather are designated np1 and np3 according to whether they
converge to the $N^+=1$ or 3 rotational level of the ion. The P(1) and
Q(1) series are pure σ and π states, respectively, and are not
affected by $\Delta v=0$ interactions. They may only be perturbed by P(1) and
Q(1) states with low principal quantum number belonging to Rydberg
series converging to higher vibrational levels of the ion ($\Delta v \neq 0$
interactions).

Figure 18 shows a schematic diagram illustrating the transitions
from the X $^1\Sigma_g^+$ state to the E,F $^1\Sigma_g^+$ state, the transitions from the
E,F $^1\Sigma_g^+$ state to the autoionizing $np\sigma$ $^1\Sigma_u^+$ and $np\pi$ $^1\Pi_u$ Rydberg states,
and the transformation from Hund's case (b) to Hund's case (d). For
molecules initially in the E,F $^1\Sigma_g^+$, v'=2, J'=1 state, only np
transitions ($\Delta\ell=\pm1$) to Rydberg states having J=0,1,2 are allowed
($\Delta J=0,\pm1$). The series must converge to odd rotational levels of the
H_2^+ core due to parity considerations. An examination of Figure 18
shows that three Rydberg series with J=0,1,2 converge to the $N^+=1$
level of the ion (corresponding to P(1)$np\sigma$, Q(1)$np\pi$, and R(1)np1
transitions) and one Rydberg series with J=2 converges to the $N^+=3$
level of the ion (corresponding to an R(1)np3 transition).

The initial step in the determination of the two color REMPI
spectrum was the measurement of the (2+1) single color ionization
spectrum of the E,F $^1\Sigma_g^+$ ← X $^1\Sigma_g^+$ transition. Figure 19 shows a small
portion of this spectrum obtained by monitoring the $v^+=1$ photoelectron
peak. This energy region contains transitions from X $^1\Sigma_g^+$, v"=0 to

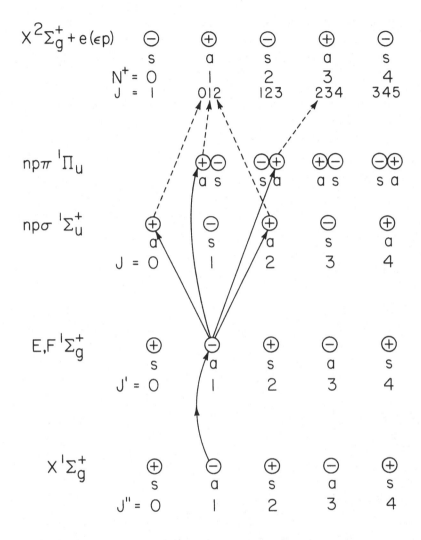

Figure 18. Schematic diagram illustrating the allowed rotational transitions for (2+1') ionization of H_2 via the E,F $^1\Sigma_g^+$ state. Processes of interest to the present experiment are shown, including: (1) the two photon transition from the X $^1\Sigma_g^+$ ground state to the E,F $^1\Sigma_g^+$ state; (2) the single photon transition from the E,F $^1\Sigma_g^+$ state to the $np\sigma$ $^1\Sigma_u^+$ and $np\pi$ $^1\Pi_u$ Rydberg states shown in Hund's case (b) coupling; and (3) the transition from Hund's case (b) to Hund's case (d) coupling.

E,F $^1\Sigma_g^+$, v'=E2 and F4. Only Q branch transitions are observed in the energy region shown in Figure 19; however, O and S branch transitions also are allowed and have been previously observed in the two photon excitation spectrum.[112-114]

It is interesting to note that the single color photoelectron spectrum shows a double peaked distribution similar to that observed previously by Anderson et al.[115] The largest peak corresponds to the formation of $v^+=2$, as expected for ionization of the Rydberg-like v'=E2 level of the inner well of the E,F $^1\Sigma_g^+$ state. However, there is also evidence of a secondary peak in the photoelectron distribution at much higher values of v^+. This secondary peak in the photoelectron

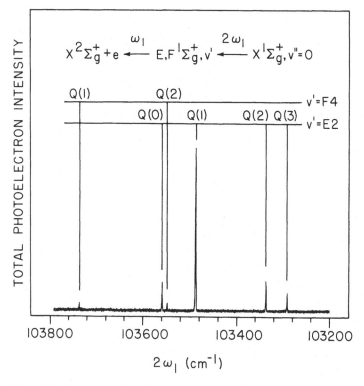

Figure 19. Single-color (2+1) excitation spectrum of a small portion of the E,F $^1\Sigma_g^+$, v'=E2 and F4 ← X $^1\Sigma_g^+$, v''=0 transition obtained by monitoring the total photoelectron signal.

distribution is due to the portion of the wavefunction of the $v'=E2$ state that is in the outer well of the E,F $^1\Sigma_g^+$ state.

Following the determination of the two photon E,F $^1\Sigma_g^+$, $v'=E2$ ← X $^1\Sigma_g^+$, $v''=0$ excitation spectrum, ω_1 was fixed at the frequency of the peak of the E2 Q(1) transition and ω_2 was scanned. Figure 20 shows the two color excitation spectrum from E,F $^1\Sigma_g^+$, $v'=2$, $J'=1$ to the autoionizing Rydberg states near the $v^+=2$ ionization limit obtained by monitoring the $v^+=1$ photoelectron peak as ω_2 was scanned. When the intensity of ω_2 was adjusted to avoid saturation, the photoelectron signal from direct, two-color ionization (i.e., the signal at energies between the autoionization peaks) was negligible.

In general, the Q(1)npπ transitions have narrow linewidths, since the $\Delta v=-1$ interactions of the pure π states with the direct ionization continuum are weak. The P(1)npσ transitions are significantly weaker than the Q(1)npπ or R(1)np1 transitions; however, they can be observed when the intensity of ω_2 is increased. The two series of R(1) transitions perturb one another both below and above the X $^2\Sigma_g^+$, $v^+=2$, $N^+=1$ threshold. In the region between the $N^+=1$ and 3 thresholds, the R(1)np3 transitions appear as a series of weak, broad, asymmetric peaks when observed in the $v^+=1$ decay channel (i.e., when the decay is via vibrational autoionization) and as a series of window resonances when observed in the $v^+=2$ decay channel (i.e., when the decay is via rotational autoionization).

Following the determination of the two color REMPI spectrum, ω_2 was tuned to a known position within an autoionizing resonance and the photoelectron time-of-flight spectrum was determined. In general, we found that the $v^+=0$ branching ratio is independent of the symmetry of the autoionizing level and varies from about 4-6%; however, two deviations from this trend are observed. The most striking departure from the average $v^+=0$ branching ratio occurs for the unresolved R(1)22p1 and Q(1)22pπ transitions at 128381.3 cm^{-1}. The region near these transitions is shown in greater detail in Figure 21. The top frame of this figure shows the excitation spectrum obtained by

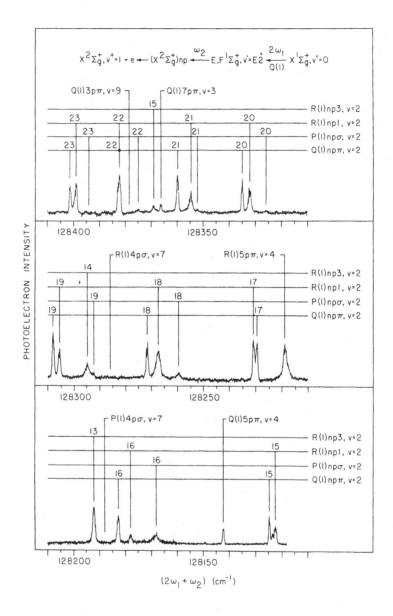

Figure 20a. Two-color (2+1') excitation spectrum of autoionizing np Rydberg states in the region of the $v^+=2$ ionization limit, obtained by monitoring the $v^+=1$ photoelectron peak. The transitions originate in the E,F $^1\Sigma_g^+$, v'=E2, J'=1 state.

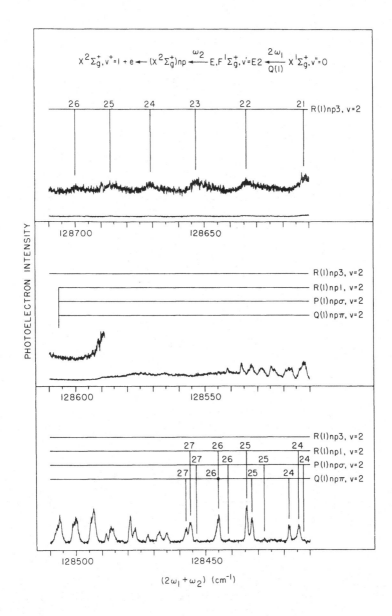

Figure 20b. Two-color (2+1') excitation spectrum of autoionizing np Rydberg states in the region of the $v^+=2$ ionization limit, obtained by monitoring the $v^+=1$ photoelectron peak. The transitions originate in the E,F $^1\Sigma_g^+$, v'=E2, J'=1 state.

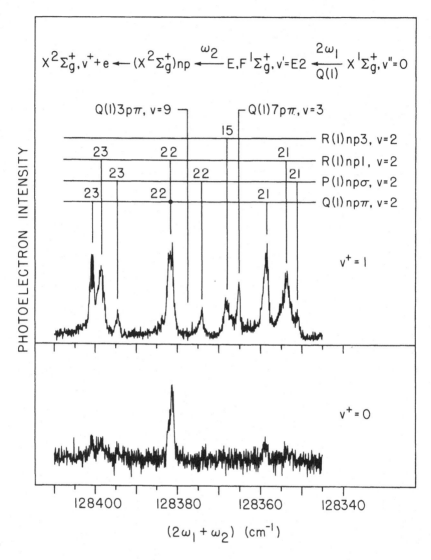

Figure 21. A small region of the two-color (2+1') excitation spectrum of autoionizing np Rydberg states shown in Figure 20. The transitions originate in the E,F $^1\Sigma_g^+$, $v'=E2$, $J'=1$ state. The top frame shows the excitation spectrum obtained by monitoring the $v^+=1$ photoelectron peak (corresponding to vibrational autoionization via $|\Delta v| \geqslant 1$) and the bottom frame shows the excitation spectrum obtained by monitoring the $v^+=0$ photoelectron peak (corresponding to vibrational autoionization via $|\Delta v| \geqslant 2$).

monitoring the $v^+=1$ photoelectron peak (corresponding to vibrational autoionization via $|\Delta v| > 1$), and the bottom frame shows the excitation spectrum obtained by monitoring the $v^+=0$ photoelectron peak (corresponding to vibrational autoionization via $|\Delta|v > 2$). These spectra were taken with a relatively high intensity of ω_2, so that the more intense lines are power broadened. This region of the spectrum contains the R(1)21-23np1, P(1)21-23pσ, Q(1)21-23pπ, and R(1)15p3 transitions, all of which converge to $v^+=2$; in addition, this region contains the Q(1)3pπ, v=9 and Q(1)7pπ, v=3 transitions. Since the autoionization rate of the 3pπ, v=9 state is very small (as it must autoionize via Δv=-8), no ionization is observed at the position of the absorption line. In general, such low n/high v interlopers will have significantly reduced autoionization rates, since autoionization must proceed via a large change in $|\Delta v|$. The excitation spectrum obtained by monitoring the $v^+=0$ photoelectron peak shows a dramatic increase near the unresolved R(1)22p1 and Q(1)22pπ transitions. A $v^+=0$ branching ratio of 17% is determined on the low energy side of this pair, decreasing to 10% at the peak, and further decreasing to 3% on the high energy side. Since the upper levels of the Q(1)22pπ and Q(1)3pπ, v=9 transitions perturb one another, it is probable that this perturbation also affects the $v^+=0$ branching ratio. The present observation would then suggest that the Q(1)22pπ transition is on the low energy side of the unresolved pair of lines.

A second example of an anomalously high $v^+=0$ branching ratio occurs for the peak at 128304.9 cm^{-1}, corresponding to the R(1)19p1 transition. A $v^+=0$ branching ratio of 10% is obtained at the peak, increasing to about 15% on the low energy side of the peak. The interaction between the upper levels of the R(1)4pσ, v=7 transition at 128286.3 cm^{-1} (which, like the 3pπ, v=9 state, is not observed in the present ionization spectrum) and the R(1)19p1 transition is the probable cause of the unusually large $v^+=0$ branching ratio.

These two deviations from the average $v^+=0$ branching ratio of 4-6% occur for Rydberg states that may be strongly perturbed by low

n/high v interlopers, i.e., the 3pπ, v=9 state and the 4pσ, v=7
state. Although transitions to these interlopers are not observed in
the present excitation spectrum, their influence is apparently felt by
nearby states. This is consistent with the results of Jungen and
Raoult,[103] who showed that interchannel coupling can spread the
oscillator strength associated with a low n/high v interloper over
several nearby Rydberg states with high n/low v. Furthermore, the
increase in the v^+=0 branching ratio for states that have some
component of low n/high v character is also consistent with the
results of Dehmer and Chupka.[111] Their work suggested that
autoionization proceeds predominantly (>95%) via the Δv=-1 process
when it is energetically allowed however, when the Δv=-1 process is
energetically forbidden, autoionization via the minimum change in Δv
is less favorable (>75%). Quantitative multichannel QDT calculations
for ionization via the E,F $^1\Sigma_g^+$, v'=E2 intermediate state are required
to confirm these arguments.

3.3.2 Competition Between Rotational and Vibrational Autoionization

In the energy region below the X $^2\Sigma_g^+$, v^+=2, N^+=1 threshold, the
dominant ionization mechanism is vibrational autoionization. Owing to
the good Franck-Condon overlap between the E,F $^1\Sigma_g^+$, v'=E2 state and
the H_2^+ X $^2\Sigma_g^+$, v^+=2 state, the direct ionization cross section into the
X $^2\Sigma_g^+$, v^+=2, N^+=1 continuum is quite large. Above the X $^2\Sigma_g^+$, v^+=2,
N^+=1 threshold, members of the R(1) np3 series can vibrationally
autoionize into the v^+=0 and 1 continua and can rotationally
autoionize into the v^+=2, N^+=1 continuum. The relative importance of
rotational and vibrational autoionization has never been determined
directly, and it is of great interest to do so for H_2, as theoretical
calculations can be readily performed.

We have recently determined the two color photoelectron spectra
of H_2 obtained by pumping the E,F $^1\Sigma_g^+$, v'=E2 Q(1) transition and
probing the region between the H_2^+ X $^2\Sigma_g^+$, v^+=2, N^+=1 and 3
thresholds. Spectra have been obtained by separately monitoring the

X $^2\Sigma_g^+$, $v^+=1$ photoelectron peak (vibrational autoionization) and the spectrum obtained by monitoring the X $^2\Sigma_g^+$, $v^+=2$, $N^+=1$ peak (rotational autoionization). The signal corresponding to the production of H_2^+ $^2\Sigma_g^+$, $v^+=0$ is too weak to be recorded with good counting statistics. As expected, the spectrum corresponding to rotational autoionization shows a sharp rise as the $v^+=2$, $N^+=1$ ionization continuum becomes energetically accessible; a corresponding drop in the spectrum corresponding to vibrational autoionization is also observed. The good Franck-Condon overlap between the E,F $^1\Sigma_g^+$, $v'=E2$ level and the X $^2\Sigma_g^+$, $v^+=2$ continuum produces a large direct ionization cross section into the X $^2\Sigma_g^+$, $v^+=2$ continuum. At shorter wavelengths, the $v^+=2$, $N^+=1$ spectrum shows a series of window resonances converging to the $N^+=3$ limit, with minima occurring at approximately the positions expected for the R(1)np3 series. The window resonances are a result of the R(1)np3 Rydberg series (which has a very small oscillator strength) interacting with the $v^+=2$, $N^+=1$ continuum (which has a relatively large oscillator strength).

The $v^+=1$ spectrum displays behavior opposite to that of the $v^+=2$, $N^+=1$ spectrum. The R(1)np3 series appears as weak asymmetric peaks, with the maxima occurring at the minima of the $v^+=2$, $N^+=1$ spectrum. It should be noted that the $v^+=2$, $N^+=1$ spectrum is more intense than the $v^+=1$ spectrum and that the total relative cross section in the region of these resonances should look much like the $v^+=2$, $N^+=1$ spectrum. Thus, the maxima in the $v^+=1$ vibrational autoionization cross section occur at the minima of the total cross section. While the appearance of these spectra is quite striking, the behavior is not unexpected. In fact, Cornaggia et al.[49] have recently performed MQDT calculations for ionization via the E,F $^1\Sigma_g^+$, $v'=E1$, $J'=1$ state, in which the ionizing photon accesses the region between the X $^2\Sigma_g^+$, $v^+=1$, $N^+=1$ and 3 limits. Levels in this region may decay by vibrational autoionization to X $^2\Sigma_g^+$, $v^+=0$ or by rotational autoionization to X $^2\Sigma_g^+$, $v^+=2$, $N^+=1$. Although this is not identical to the present study, the situations are strictly analogous. Calculations of the total cross section[49] predict a series of window resonances with the same asymmetries as those observed in the experimental spectra. In addition, the branching ratio into $v^+=0$ also shows small, asymmetric peaks, analogous to those observed in the $v^+=1$ spectrum. The remarkably good agreement between experiment and theory indicates that we have now reached the point where we can compare experiment and theory for photoionization processes in H_2 with unprecedented detail.

3.4 REMPI-PES Studies of Autoionizing States of NO

Several studies analogous to those on H_2 discussed above have been performed on NO. The combination of REMPI-PES and two-color excitation via the A $^2\Sigma^+$ or C $^2\Pi$ states to high Rydberg states with vibrationally excited ion cores allows the detailed study of vibrational autoionization of these levels.[116-118] In some instances, the vibrational branching ratios deviate significantly from the $\Delta v=-1$ propensity rule discussed above for "pure" vibrational autoionization.[116,118] These deviations are discussed in terms of the MQDT analysis of NO performed by Giusti-Suzor and Jungen.[119] In their analysis, the competition between autoionization and predissociation is taken into account, and is found to produce significant breakdowns of the propensity rule. Although calculations for the specific processes of the REMPI-PES studies have not been made, the experimental data strongly support this explanation.

In another MQDT study, Fredin et al.[120] have performed a detailed analysis of their two-color REMPI spectra obtained by pumping the C $^2\Pi$, v=0 state and probing the ns and nd v=0 Rydberg states. In addition to providing a unified description of these series from low n to very high n, Fredin et al.[120] have also performed a calculation of the two-color, zero kinetic energy photoelectron spectra of Müller-Dethlefs et al.[25] The excellent agreement indicates that the MQDT analysis contains all of the essential physics of this problem.

We have recently used one color, two photon excitation to perform a REMPI-PES study of the autoionization dynamics of the rotationally resolved NO $9d\pi^-$, v=2 state, which can decay into both the NO^+ X $^1\Sigma_u^+$, $v^+=0$ and $v^+=1$ continua.[121] Figure 22 shows the REMPI spectrum of this band and Figure 23 shows a typical photoelectron spectrum of NO obtained in this region with the analyzer parallel to the polarization axis of the light. Although the relative intensities will be discussed more quantitatively below, the $v^+=1$ peak is significantly larger than the $v^+=0$ peak, as is expected on the basis of Franck-Condon factors between the NO X $^2\Pi_{1/2}$, v=0 and the NO^+ X $^1\Sigma^+$, v^+ states. Photoelectron spectra similar to that of Figure 23 were obtained as a function of wavelength as the laser was scanned across three different autoionizing resonances in Figure 22. The data were

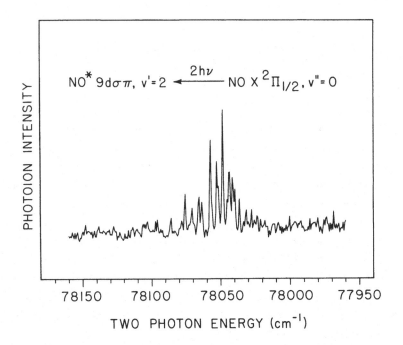

Figure 22. The two photon ionization spectrum of supersonically cooled NO between 2655 and 2665 Å. The abscissa gives the two photon energy in cm^{-1}.

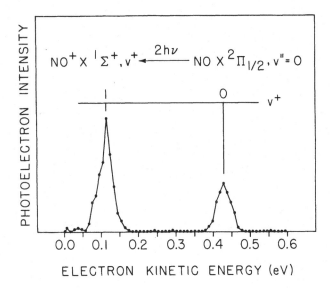

Figure 23. The photoelectron spectrum of NO following two photon ionization at 2562.43 Å (78050.9 cm^{-1}) and with the polarization axis of the laser beam parallel to the detector axis (θ=0°).

plotted by integrating the v^+=0 and v^+=1 photoelectron peak intensities and plotting the intensity ratio (at θ=0°) into the two vibrational levels of NO$^+$ as a function of the two photon energy. The portion of the data containing the autoionizing resonance at 78048.4 cm^{-1} is shown in Figure 24. The θ=0° branching ratio into the NO$^+$ $^1\Sigma^+$, v^+=1 state shows a pronounced increase at the autoionizing resonance, with a corresponding decrease in the intensity ratio into the v^+=0 state. This behavior is also observed for the neighboring resonances, and is consistent with expectations based on the $\Delta v = -1$ propensity rule. Thus, on the autoionizing resonance, the mechanism of vibrational autoionization enhances the v^+=1 photoelectron peak relative to the v^+=0 peak.

However, because the two photoelectron peaks may have different photoelectron angular distributions, the relative peak intensities at θ=0° may not reflect the true angle integrated branching ratio (i.e.,

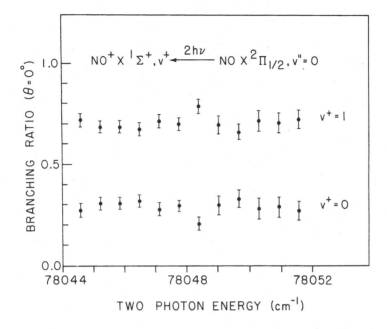

Figure 24. The vibrational intensity ratios at θ=0° following two photon ionization of NO into the $NO^+ X \, ^1\Sigma_g^+$, $v^+=0,1$ continua as a function of energy. A resonance occurs at 78048.4 cm^{-1}.

the relative partial cross section). For this reason, the photoelectron angular distributions were determined at four wavelengths across the band, including the peak of the autoionizing resonance (78048.4 cm^{-1}), just off the same resonance (78047.1 cm^{-1}), and at two off resonance positions (78080.1 and 78143.6 cm^{-1}). The four photoelectron angular distributions for the $v^+=1$ peaks are shown in Figure 25. Qualitatively, the most striking feature is the difference between the on-resonance spectrum and the three off-resonance spectra, the former being significantly more isotropic. This indicates that on resonance, in the present case, anisotropy is being transferred from the photoelectron to the photoion, resulting in a more isotropic electron angular distribution. Of course, it should be stressed that the opposite can happen, depending on the particular dynamics of the autoionizing state.

In the electric dipole approximation the photoelectron angular distribution for two photon ionization of an unaligned sample with linear polarized light has the function form[22,122,123]

$$I_{v^+}(\theta) \propto 1 + \beta_{v^+}P_2(\cos\theta) + \gamma_{v^+}P_4(\cos\theta). \tag{3}$$

Here, $I_{v^+}(\theta)$ is the photoelectron intensity, θ is the angle between the polarization axis of the light and the detector, β_{v^+} and γ_{v^+} are asymmetry parameters, and $P_2(\cos\theta)$ and $P_4(\cos\theta)$ are the second and fourth Lengendre polynomials, respectively $[P_2(\cos\theta) = 1/2(3\cos^2\theta-1);$ $P_4(\cos\theta) = 1/8(35\cos^4(\theta) - 30\cos^2(\theta) + 3)]$. The values of β_{v^+} and γ_{v^+} have been determined for the four distributions using a least squares fit to the data. These fits are shown in Figure 25 and are summarized in Table 5. Also shown in Figure 25 are the best fits obtained by ignoring the contribution of the P_4 term, and it is clear that the P_4 term must be included to get a reasonable fit to the data. The best fit values of β_{v^+} and γ_{v^+} in Table 5 reinforce the observed differences between the on-resonance and off-resonance angular distributions. In addition, there appears to be a smaller increase in the value of β_{v^+} as the two photon energy becomes further off-resonance.

As discussed above, the photoelectron angular distributions for the $v^+=0$ peak were determined by recording the full photoelectron spectrum at five angles and scaling to the $v^+=1$ result. The $v^+=0$ angular distributions for the three off-resonance energies were identical to the corresponding $v^+=1$ angular distribution to within the experimental uncertainty. Only the on-resonance $v=0$ distribution showed a significant difference, falling off much faster at larger angles than the $v^+=1$ distribution, as is seen in Figure 26. A fit to the (albeit limited) data gives $\beta = 1.3 \pm 0.2$ and $\gamma = 0.8 \pm 0.2$ ($\beta = 1.8 \pm 0.2$ if the P_4 term is ignored). These values are much closer to the off-resonance $v^+=1$ parameters than to the on-resonance $v^+=1$ parameters.

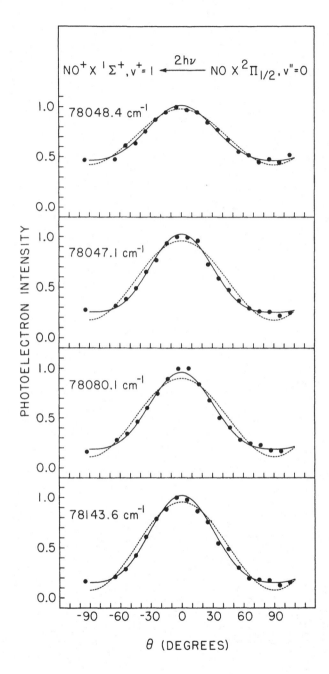

Figure 25. The photoelectron angular distributions following two photon ionization of NO for the NO$^+$ X $^1\Sigma^+$, v$^+$=1 peak at four different energies. The experimental data are given as solid circles (●). The solid line (——) represents a best fit to the function form 1 + $\beta P_2(\cos\theta)$ + $\gamma P_4(\cos\theta)$ and the dashed line (- - -) represents a best fit to the functional form 1 + $\beta P_2(\cos\theta)$.

Table 5. Angular distribution parameters and angle integrated
branching ratios for two photon ionization of NO.

Two photon energy (cm^{-1})	v^+	β_{v^+}	γ_{v^+}	$\dfrac{\sigma_v^+}{\sigma_0 + \sigma_1}$
78048.4	0	1.3 ± 0.2	0.8 ± 0.2	0.11
(on-resonance)	1	0.6 ± 0.1	0.1 ± 0.1	0.89
78047.1	0	1.1 ± 0.2	0.4 ± 0.2	0.26
	1	1.1 ± 0.1	0.4 ± 0.1	0.74
78080.1	0	1.3 ± 0.2	0.4 ± 0.2	0.29
	1	1.3 ± 0.1	0.4 ± 0.1	0.71
74143.6	0	1.4 ± 0.2	0.4 ± 0.2	0.29
	1	1.4 ± 0.1	0.4 ± 0.1	0.71

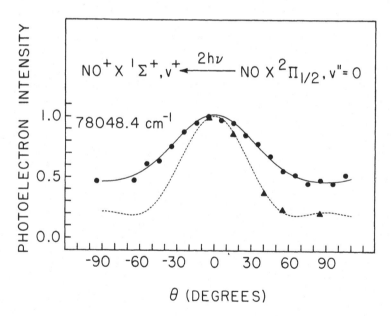

Figure 26. The photoelectron angular distributions following two
photon ionization of NO for the NO$^+$ X $^1\Sigma^+$, v^+=0 and v^+=1 peaks. The
experimental data are given as solid triangles (▲) for v^+=0 and solid
circles (●) for v^+=1. The dashed line (- - -) and solid line (⸺)
represent best fits to the functional form $1 + \beta P_2(\cos\theta) + \gamma P_4(\cos\theta)$
for v^+=0 and 1, respectively.

The angle integrated branching ratios can be calculated from the $\theta = 0°$ intensity ratios and the photoelectron angular distributions using the formula

$$\frac{\sigma_{v^+}}{4\pi} = \frac{d\sigma_{v^+}/d\Omega}{1 + \beta_{v^+}P_2(\cos\theta) + \gamma_{v^+}P_4(\cos\theta)} , \tag{4}$$

where σ_{v^+} and $d\sigma_{v^+}/d\Omega$ are the total cross section and the differential cross section into the v^+ ionic state, respectively. The measured intensity $I_{v^+}(\theta)$ is proportional to $d\sigma_{v^+}/d\Omega$, and, assuming the detection efficiency of the apparatus is independent of v^+, the proportionality constant C is also independent of v^+. At $\theta=0°$, Equation (3) reduces to

$$\frac{\sigma_{v^+}}{4\pi} = \frac{(d\sigma_{v^+}/d\Omega)(0^U)}{1 + \beta_{v^+} + \gamma_{v^+}} = \frac{CI_{v^+}(0^U)}{1 + \beta_{v^+} + \gamma_{v^+}} , \tag{5}$$

and the angle integrated branching ratios are equal to

$$\frac{\sigma_{v^+}}{\sigma_{v^+=0} + \sigma_{v=1}} = \frac{I_{v^+}(0^U)}{1 + \beta_{v^+} + \gamma_{v^+}} \left[\frac{I_0(0^U)}{1 + \beta_0 + \gamma_0} + \frac{I_1(0^U)}{1 + \beta_1 + \gamma_1} \right]^{-1}. \tag{6}$$

Because the $v^+=0$ and $v^+=1$ angular distributions are identical for the off-resonance energies, the angle integrated branching ratios for these energies are equal to the $\theta=0°$ values.

The branching ratios at the four energies of the present study are summarized in Table 5. As expected from a visual inspection of Figures 23 and 24, the branching ratio into the $v^+=1$ state increases dramatically for the on-resonance wavelength, particularly in view of the large nonresonant contribution (see below) and the finite wave-length resolution, which will have the effect of reducing the resonance induced variations. The photoelectron spectra have not been

corrected for the transmission function of the electron spectrometer. While this correction is not expected to be large, the
possible error introduced into the present results is discussed
here. First, the transmission function will be the same for all of
the wavelengths of the present study, so that qualitative changes in
the spectra as a function of wavelength are not affected. However, in
principle, the relative magnitudes of the $v^+=0$ and 1 peaks need to be
corrected in each of the spectra. While this cannot affect the
qualitative observation that the branching ratio into the $v^+=1$ channel
increases on resonance, the quantitative values of the branching
ratios in Table 5 would be affected. One estimate of the magnitude of
the possible error can be obtained by a comparison of the off-
resonance branching ratios with the single photon result of Ono et
al.[124] and with the results of Franck-Condon factor calculations.[125]
The ratio of the $v^+=0$ to $v^+=1$ branching ratios range from 0.56 to 1
for the single photon experimental result[124] to 0.478 to 1 for the
calculations,[125] in fair agreement with the 0.43 to 1 result for the
present data. Although this comparison provides some indication of
the magnitude of the possible error, we caution that the comparison of
Franck-Condon factors for one and two photon processes is not
necessarily straightforward, due to the summation over all possible
intermediate levels in the latter case.

Although, strictly speaking, the two processes cannot be
separated, it is interesting to estimate the contributions of direct
ionization and autoionization to the overall vibrational branching
ratios determined at the energy of the autoionizing resonance. The
direct ionization continuum is approximately 28% of the total
intensity of the resonance and using the off-resonance branching ratio
for the contribution from direct ionization

$$0.28\ I_{v^+=1}\ \text{(direct ionization)} + 0.72\ I_{v^+=1}\ \text{(autoionization)} =$$
$$I_{v^+=1}\ \text{(on-resonance),}$$

$$0.28(0.71) + 0.72\ I_{v^+=1}\ \text{(autoionization)} = 0.89,$$

from which one finds the autoionization branching ratio into $v^+=1$ is ~0.96. It seems likely that if the autoionizing resonance could be accessed from a Rydberg level with an NO^+ X $^1\Sigma^+$, $v^+=2$ ion core, as in an optical-optical double resonance experiment, the direct ionization continuum would be extremely weak due to the small Franck-Condon overlap, and the autoionizing resonance would produce essentially pure NO^+ X $^1\Sigma^+$, $v^+=1$.

3.5 Circular Dichroism of Photoelectron Angular Distributions

It is well known that the photoelectron angular distribution following single photon ionization of an unaligned sample using linearly polarized light must have the functional form[126]

$$I(\theta) \propto 1 + \beta_2 P_2(\cos\theta). \tag{7}$$

Here I is the photoelectron intensity, θ is the angle between the polarization axis of the light and the detector, β_2 is an asymmetry parameter, and P_2 is the second Legendre polynomial. For (m+n) ionization of an unaligned sample, the angular distribution may include higher order Legendre polynomials, and in general has the form[22,122,123]

$$I(\theta) \propto \sum_{i=0}^{K} A_{2i} P_{2i}(\cos\theta), \tag{8}$$

where K is the total number of photons (i.e. m+n). This is a generalization of Equation (3) above. If the (m+n) resonant ionization process is considered in two parts, the first m photons can be thought of as producing an aligned excited state, while the ionization process corresonds to n photon ionization of an aligned sample. Alignment is defined in a broad sense as a non-statistical M_J population distribution for a given J in the excited state. A full understanding of the dynamics of REMPI therefore requires a determination of the alignment of the resonant intermediate level, which, in general, cannot be determined by the photoelectron angular distribution alone. One technique that provides some information on the alignment of the resonant intermediate level involves measuring the polarization of fluorescence from the excited state, or equivalently, the fluorescence angular distribution.[127] While this technique has been demonstrated in REMPI studies of NO[128] and has been extremely useful in a number of other applications,[127] it suffers from several limitations and is not applicable to resonant intermediate levels that do not fluoresce.

Recently Dubs et al.[33,129-131] have developed the theoretical basis of a new probe of alignment that involves the determination of the circular dichroism of photoelectron angular distributions (CDAD). The circular dichroism signal corresponds to the difference between the photoelectron angular distributions obtained using left and right hand circularly polarized light. The CDAD signal must be zero for symmetry reasons at $\theta=0°$ and $90°$; for an unaligned sample, the CDAD signal is zero for all angles. However, with the appropriate detection geometry, alignment of the sample breaks the cylindrical symmetry of the system. In this case the system responds differently to left and right hand circularly polarized light, giving rise to the CDAD signal.

The first experimental CDAD study was performed by Appling et al.[132] on NO. In these experiments, linearly polarized light was used to pump a two photon transition from the X $^2\Pi_{1/2}$, v"=0 ground state to the A $^2\Sigma^+$, v'=0 state, and the alignment produced by this transition was probed using counterpropagating left and right hand circularly polarized light to ionize the A $^2\Sigma^+$, v'=0 state. The electrons were detected perpendicular to the direction of propagation of the light. The experimental CDAD spectra show significant dichroic effects, which are approximately 20% of the differential cross section in magnitude. They are also in reasonably good agreement with the theoretical calculations of Dubs et al.[33] This indicates that the theoretical framework and calculations provide a realistic description of the photoionization dynamics.

Dubs et al. have recently shown how CDAD spectra can be analyzed to extract the moments of alignment, at least in principle, to any order.[131] They have also shown how CDAD measurements can be used to probe alignment in photodissociation products.[130] A significant advantage of CDAD over laser induced fluorescence techniques is that it can be applied to molecules that do not fluoresce. It should also be stressed that while photoelectron angular distributions following REMPI contain information on both the alignment of the intermediate state and the anisotropy of the ionizing transition, these two aspects of the problem are generally not easy to unravel. The corresponding CDAD spectra contain information only on the alignment of the resonant intermediate state, and therefore may aid in the understanding of photoelectron angular distributions following REMPI. Although the ultimate utility of CDAD techniques must await further experimental studies, it represents an interesting adaptation of REMPI-PES to a different area of research.

4. PHOTOELECTRON SPECTROSCOPY OF VAN DER WAALS MOLECULES AND FREE RADICALS

4.1 REMPI-PES of KrXe

In the past few years REMPI-PES has proven to be an effective technique for the determination of photoelectron spectra of van der Waals molecules.[27,133-138] As discussed above, the selective ionization provided by REMPI is particularly important for the study of van der Waals molecules since they are usually a minor component of the molecular beam. As will be shown below, REMPI-PES is also a useful tool for the study of the predissociation dynamics of the resonant intermediate level. In the discussion that follows, we will focus on our work on the rare gas van der Waals dimers,[135-138] although it should be noted that a number of other systems have been studied by other groups using REMPI-PES.[27,133,134]

Several years ago, photoelectron spectra of the homonuclear rare gas van der Waals dimers Ar_2, Kr_2, and Xe_2 were determined using conventional HeI photoelectron spectroscopy.[14-16] More recently, photoelectron-photoion coincidence spectroscopy was used to obtain the photoelectron spectrum of the xenon trimer.[139,140] However, neither technique has succeeded in producing an unambiguous photoelectron spectrum for any of the heteronuclear rare gas dimers. The primary difficulty in using either of these techniques to determine the heteronuclear dimer spectra is interference from ionization of the corresponding free atoms and homonuclear rare gas dimers. The selective ionization that is possible using REMPI removes this difficulty, as will be discussed for the case of KrXe, below. REMPI-PES has also been used to record the photoelectron spectra of the heteronuclear rare gas dimer ArXe.[136]

The present experiments were performed using a cw unskimmed supersonic expansion. An 8:1 mixture of Kr:Xe was expanded through a 12.5 μm jet with a stagnation pressure of 2.72 atm. which produces a concentration of approximately 1% KrXe dimers in the beam with a negligible concentration of larger heteronuclear clusters. Because the spectroscopy of KrXe is not well known, the experiments were performed in two steps. First, the REMPI spectrum was recorded in the region of interest by monitoring the KrXe$^+$ ion signal using a time-of-flight mass spectrometer while the laser wavelength was scanned. The laser was then tuned to the KrXe feature of interest and the REMPI-PES spectrum was recorded. In these experiments, the (2+1) ionization spectra of KrXe$^+$ were recorded in the region of the atomic Xe 6p transitions. As an example, the spectrum in the region of the atomic Xe 6p[5/2]$_2$ level[141] at 78120.30 cm^{-1} is shown in Figure 27. The rotational structure is very dense and is not resolved with the present resolution. The bands of Figure 27 have not been observed previously, and their analysis provides information on the potential curves of the resonant intermediate state. The analysis of REMPI mass spectra has grown into a large field[142] and is beyond the scope of this chapter.

The REMPI-PES spectrum of KrXe obtained at the wavelength of the KrXe (0,0) band of Figure 27 is shown in Figure 28. No signal is

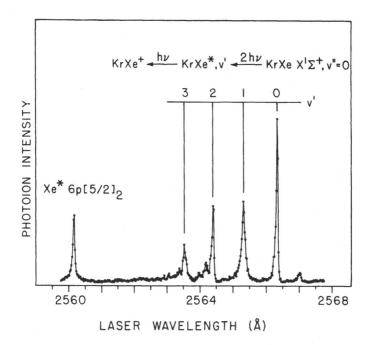

Figure 27. The (2+1) REMPI spectrum of KrXe obtained by monitoring the KrXe[+] ion signal. The peak at 2560.154 Å results from (2+1) ionization of atomic Xe via the $6p[5/2]_2$ state. Ion signal appears at the mass of KrXe[+] as a result of tailing of the very intense Xe[+] mass peak, and is used to calibrate the spectrum.

observed in the unplotted regions between 0.0 and 0.9 eV and between 1.25 and 2.10 eV. The spectrum of Figure 28 displays five peaks, with the highest kinetic energy peak being the most intense. It is seen that there is little or no photoelectron intensity at kinetic energies corresponding to non-resonant, three photon ionization of atomic Xe, indicating that this process is very weak at the wavelength of the KrXe (0,0) band. The most intense peak in Figure 28 appears at a kinetic energy of 2.438 eV, which is greater than that expected for non-resonant photoionization of atomic Xe, indicating that the peak results from photoionization of KrXe[*] to produce a bound state of

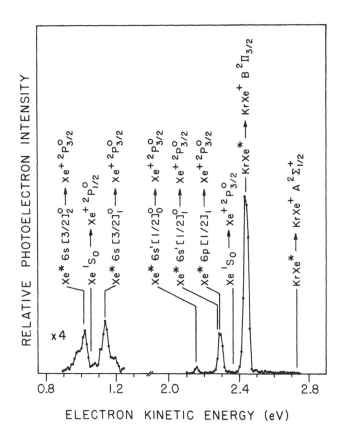

Figure 28. Photoelectron spectrum of KrXe following (2+1) ionization at 2566.33 Å. The calculated positions of photoelectron peaks resulting from (three-photon) non-resonant ionization of ground state xenon atoms and from (single-photon) ionization of excited xenon atoms produced by predissociation of the KrXe* resonant intermediate level are indicated. The adiabatic ionization potential of KrXe,[143] which corresponds to a photoelectron kinetic energy of 2.731 eV, is also indicated.

$KrXe^+$. This bound state corresponds to one of only two states of $KrXe^+$ that dissociates to $Kr + Xe^+\ ^2P^o_{3/2}$; these states are the A $^2\Sigma^+_{1/2}$ ground state, which has a dissociation energy of 0.385 eV,[143] and the B $^2\Pi_{3/2}$ excited state, which is weakly bound, although no experimental value for the dissociation energy has been determined.

These results provide information on both the electronic character of the resonant intermediate state and the ionic state. For example, REMPI via the atomic Xe^* $6p[5/2]_2$ level (to which the molecular resonant intermediate level dissociates) produces essentially pure Xe^+ $^2P^o_{3/2}$, showing that the neutral atomic level has the $^2P^o_{3/2}$ ion core. This, together with the very small dissociation energy (0.042 eV) of the resonant intermediate level, indicates that the $KrXe^*$ state has the $KrXe^+$ B $^2\Pi_{3/2}$ ion core. If the electronic state of the ion core is preserved upon ionization, as has been discussed above,[60] the photoelectron peak at 2.438 eV in Figure 28 can be assigned to the production of $KrXe^+$ in the B $^2\Pi_{3/2}$ state. This assignment is supported by a qualitative consideration of the Franck-Condon factors for the ionizing transition. These Franck-Condon factors show that ionization of the weakly bound resonant intermediate level to the relatively strongly bound $KrXe^+$ A $^2\Sigma^+_{1/2}$ electronic state should populate a broad distribution of final vibrational states. Because the vibrational frequency of the A $^2\Sigma^+_{1/2}$ state is expected to be smaller than the electron spectrometer resolution, this would lead to a broad photoelectron peak, which is not observed. However, because the potential energy curve of the resonant intermediate state is more similar to that of the $KrXe^+$ B $^2\Pi_{3/2}$ electronic state, the Franck-Condon factors for this ionizing transition should be more nearly diagonal. This should produce a narrower photoelectron peak, which is observed. Thus, the photoelectron peak is assigned to the $KrXe^+$ B $^2\Pi_{3/2}$ state.

The kinetic energy of the photoelectron peak can then be used to provide a lower limit to the dissociation energy of the B $^2\Pi_{3/2}$ state, which is given by $D_0(KrXe^+$ B $^2\Pi_{3/2}) > IP(Xe^+$ $^2P^o_{3/2}) - IP(KrXe^+$ B $^2\Pi_{3/2}) + D_0(KrXe)$. Here, $IP(Xe^+$ $^2P^o_{3/2})$ is the ionization potential of atomic Xe (12.130 eV),[141] $IP(KrXe^+$ B $^2\Pi_{3/2})$ is the ionization potential of $KrXe^+$, which is equal to the (three-) photon excitation energy (14.494 eV) minus the kinetic energy of the photoelectron peak (2.438 eV); and $D_0(KrXe)$ is the dissociation energy of the neutral KrXe ground state (0.018 eV).[144] Thus,

$D_0(\text{KrXe}^+ \ B \ ^2\Pi_{3/2}) \geq 0.092 \pm 0.006$ eV. By suitable choice of the resonant intermediate state, it should be possible to study the other ionic states of KrXe using REMPI-PES. For example REMPI via KrXe* levels that dissociate to Kr + Xe* 6p', that is, excited Xe atoms with $^2P^o_{1/2}$ ion cores, would be expected to produce predominantly KrXe$^+$ B $^2\Pi_{1/2}$, while REMPI via levels that dissociate to Kr* + Xe would allow the study of the KrXe$^+$ C $^2\Pi_{3/2}$, C $^2\Pi_{1/2}$, and D $^2\Sigma^+_{1/2}$ states.

We now show that the four remaining peaks in the photoelectron spectrum result from predissociation of the KrXe* resonant intermediate level to form Kr + Xe*, followed by single-photon ionization of the excited xenon atom. There are only five excited states of Xe* that lie below the energy of the KrXe* resonant intermediate state.[141] The kinetic energies of photoelectron peaks produced by a single-photon ionization of these states may be calculated using the relation KE = $h\nu$ + E(Xe*) – IP(Xe $^2P^o_{3/2}$), where $h\nu$ is the (single-) photon energy, E(Xe*) is the energy of the excited state of atomic Xe, and IP(Xe $^2P^o_{3/2}$) is the ionization potential of Xe. The calculated kinetic energies of these five peaks are indicated in Figure 28. The agreement between the calculated and the observed peak positions is excellent, indicating that predissociation of the resonant intermediate level is indeed the mechanism for production of the extra peaks in the photoelectron spectrum. Because the excited atomic fragments are ionized within the 5 ns laser pulse, which in most instances is shorter than both the fluorescence lifetime of the level and the mean collision time, the relative intensities of the photoelectron peaks provide information on the nascent branching ratios of the predissociation process. Qualitatively, the observation of photoelectron peaks following predissociation can be used to determine the dissociation limits of the predissociating states. Although similar information has been obtained previously on a number of systems using dispersed fluorescence to determine the electronic state of the excited fragment,[145] in some instances fluorescence may be extremely weak or forbidden. In these cases, REMPI-PES provides an alternative technique.

The quantitative determination of predissociation branching ratios is more difficult for two reasons. First, the photoionization cross section at the incident wavelength may depend on the resonant intermediate atomic state; however, under the appropriate circumstances it may be possible to saturate the ionizing transition, and therefore eliminate this difficulty. Second, the photoelectron angular distribution may depend on the resonant intermediate atomic state. Because the excited atomic fragments with $J \neq 0$ will, in general, be aligned following the predissociation process,[127] there is no magic angle at which the photoelectron spectrum can be related to the true partial cross sections, and it is necessary to determine the photoelectron spectrum at several different angles. An alternative would be to use the magnetic bottle electron spectrometer[23] discussed above. Since this spectrometer has a 50% electron collection efficiency, the relative photoelectron intensities will reflect the true partial cross sections. Thus, although some difficulties exist in extending the present measurements to provide quantitative branching ratios for predissociation, the resolution of these problems is possible.

4.2 REMPI-PES of Xe_2

The comparison of the REMPI-PES spectra of Xe_2[146] with the HeI-PES of Xe_2 provides a good example of the power of the former technique in the determination of ionic energy levels. Mulliken first discussed the general properties of the potential curves of the rare gas dimer ions using Xe_2^+ as an example.[147,148] He showed that four electronic states (A $^2\Sigma_u^+$, B $^2\Pi_g$, C $^2\Pi_u$, and D $^2\Sigma_g^+$) arise from the combination Xe $^1S_0$ + Xe^+ $^2P^o_{3/2}$ or $^2P^o_{1/2}$. Spin-orbit splitting of the B $^2\Pi_g$ and C $^2\Pi_u$ states yields a total of six electronic states. As a result of the removal of the outermost antibonding electron, the A $^2\Sigma^+_{1/2u}$ ground state is bound; however, all five excited states were predicted by Mulliken to be repulsive with only a small polarization minimum at large internuclear distance (R). At small R, the states are, in order of increasing energy, A $^2\Sigma^+_{1/2u}$, B $^2\Pi_{3/2g}$, B $^2\Pi_{1/2g}$,

C $^2\Pi_{3/2u}$, C $^2\Pi_{1/2u}$, and D $^2\Sigma^+_{1/2g}$. At large R the states are labeled according to Hund's case (c) coupling rules, and their ordering may be different, but must be consistent with the appropriate noncrossing rules. Recent calculations that include the effects of spin-orbit coupling[149,150] generally support Mulliken's predictions[147,148] of weak binding in the excited states of the dimer ions; however, the B $^2\Pi_{1/2g}$ state is calculated to be repulsive at all internuclear distances. The remaining excited states are bound by ion-induced dipole forces and have calculated dissociation energies of ~0.003–0.21 eV.[149] The calculations also predict a curve crossing between the B $^2\Pi_{1/2g}$ and C $^2\Pi_{3/2u}$ excited states at an R less than that of the potential minimum of the second; thus the ordering of the states given above is altered at larger R. Figure 29 shows a schematic potential energy level diagram for these states.

The HeI photoelectron spectrum[14,15] of Xe_2 in the region of the Xe^+ $^2P^o_{3/2}$ + Xe $^1S_0$ dissociation limit is shown in the lower frame of Figure 30. Three peaks can be assigned to Xe_2^+, although the B $^2\Pi_{3/2g}$

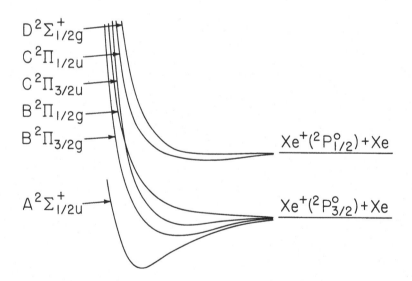

Figure 29. Schematic potential energy curves for Xe_2^+.

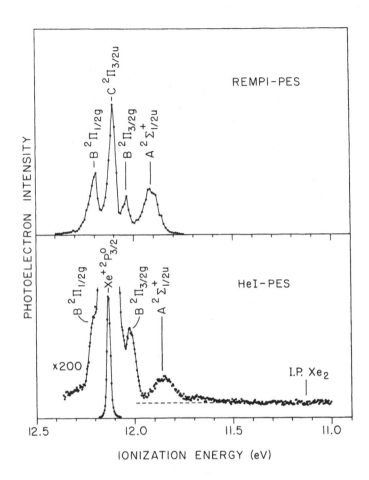

Figure 30. Illustrative example of photoelectron spectra of Xe_2 obtained by using HeI ionization (lower trace) and resonantly enhanced multiphoton ionization (upper trace) in the region of the Xe^+ $^2P^0_{3/2}$ ionization limit. Note that, in the HeI-PES, atomic ionization is more than 100 times as intense as the molecular ionization, but that no atomic ionization is observed in the REMPI-PES.

band is somewhat overlapped by the atomic Xe peak and the B $^2\Pi_{1/2g}$ band is barely observed as a shoulder of the atomic peak. In contrast, the upper frame of Figure 30 shows the photoelectron spectrum obtained following (2+1) ionization of Xe_2 via the (0,0) vibrational band of a molecular state dissociating to $Xe^* \, 5d[5/2]_2^o + Xe \, ^1S_0$. The REMPI spectrum in this region[151] is shown in Figure 31. All four of the Xe_2^+ ionic states predicted in this energy region are observed and clearly resolved in Figure 30. This is possible due to the lack of interference by the atomic Xe peak. In fact, there is no evidence for nonresonant three photon ionization of atomic Xe in any of our spectra of Xe_2 obtained so far. By recording

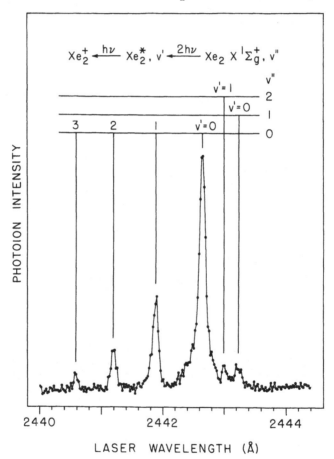

Figure 31. The (2+1) ionization spectrum of Xe_2 in the region of the atomic $Xe^* \, 5d[5/2]_2^o$ state.

REMPI-PES spectra of Xe_2 via a number of resonant intermediate levels, information has been obtained on all six dimer ion states,[146] two of which had not been observed previously. The dissociation energies determined from the REMPI-PES spectra are summarized in Table 6.

Table 6. Spectroscopic Properties for Xe_2^+

| State | D_0 (eV) | | Theory[a] | |
	Present Work	Other Experimental Determinations	D_e (eV)	R_e (Å)
A $^2\Sigma_{1/2u}^+$	0.75	1.034[b]	1.072	3.18
B $^2\Pi_{3/2g}$	0.21	0.18_5[c]	0.149	3.91
C $^2\Pi_{3/2u}$	0.05		0.057	4.28
B $^2\Pi_{1/2g}$			Repulsive	Throughout
C $^2\Pi_{1/2u}$	0.17	0.19[c]	0.211	3.97
D $^2\Sigma_{1/2g}^+$	0.04		~0.025	~4.34

[a]Reference 149. [b]Reference 152. [c]Reference 15.

Recently an analogous but more limited, REMPI-PES study was performed for Kr_2.[138]

In general, two factors require that REMPI-PES spectra be obtained via a number of resonant intermediate states in order to study all of the ionic states of interest. First, since each resonant intermediate state has a unique core (which may actually be a mixture of two or more ion states) and since core switching transitions are unfavored in the ionization step, it is not usually possible to observe all of the dimer ion bands in a single REMPI-PES. Indeed, the observation of all four molecular bands in the energy region near the Xe $^1S_0$ + Xe$^+$ $^2P_{3/2}^o$ dissociation limit in the REMPI-PES shown in Figure 30 is atypical. More often, a REMPI-PES of Xe_2 obtained in this energy region will show from one to three molecular bands. Second, as discussed above, REMPI-PES spectra may contain photoelectron peaks due to predissociation of the resonant intermediate state followed by photoionization of the excited atomic fragment, i.e. $Xe_2^* \rightarrow Xe + Xe^* \rightarrow Xe + Xe^+ + e^-$. Strongly predissociated bands will show no evidence of molecular bands in their photoelectron spectra and will not be useful for the determination of ionic properties of the dimer.

Several of the photoelectron spectra obtained for Xe_2 show evidence for partial predissociation of the resonant intermediate level, while others show no evidence for predissociation (i.e. the REMPI-PES spectrum of Figure 30).[146] A particularly striking example of the effects of predissociation is found in the Xe_2 bands observed in the region of the Xe^* $5d[7/2]_3^o$ state. The REMPI spectrum in this region[151] is shown in Figure 32. The absolute vibrational quantum number in the upper state has not been assigned.[151] Although the spectrum clearly displays molecular Xe_2 structure, it was actually obtained by monitoring the Xe^+ ion signal, rather than the Xe_2^+ ion signal. The REMPI-PES spectrum obtained for the (m,0) vibrational band is shown in Figure 33. The spectrum shows no evidence for photoionization of molecular levels, indicating that the predissociation rate is significantly higher than the photoionization rate out of this resonant intermediate level. The uncertainty in the calibration of the photoelectron energy scale and the finite electron spectrometer resolution preclude a definite assignment of the predissociation fragment; however, the REMPI-PES determined via (2+1) ionization of the atomic $6p[1/2]_0$ level shows the same Xe^+ $^2P_{3/2}^o : {}^2P_{1/2}^o$ branching ratio as that shown in Figure 33 suggesting that the resonant intermediate state predissociates solely to $6p[1/2]_0$.

4.3 REMPI-PES of Free Radicals

The use of REMPI-PES for the study of free radicals offers many of the same advantages as it does for the study of van der Waals

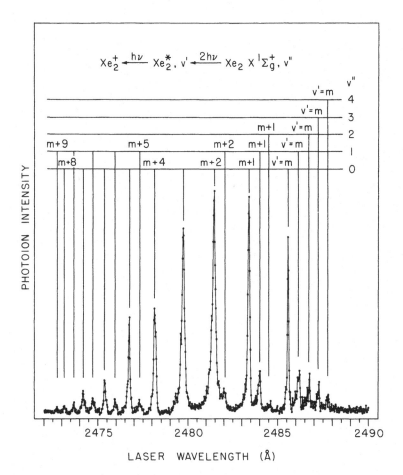

Figure 32. The (2+1) ionization spectrum of Xe_2 in the region of the atomic Xe^* $5d[7/2]_3^0$ state.

molecules. An additional advantage is that it is often possible to create the free radical of interest by photodissociation of an appropriate precursor molecule. In this manner it has been possible to use REMPI-PES to study the excited state photoionization dynamics of a number of open shell atoms, such as C,[153] N,[154] Cl,[155] Br,[24] and I,[156] as well as molecular free radicals. The use of REMPI alone to record the spectra of free radicals has been an active field and is

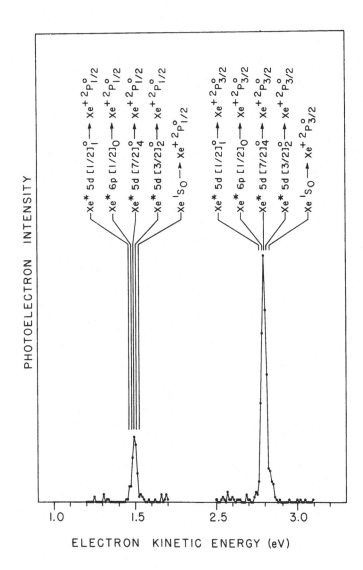

Figure 33. REMPI-PES of Xe_2 obtained via (2+1) ionization of the (m,0) vibrational band of a state dissociating to Xe $^1S_0$ + Xe* $5d[7/2]_3^o$.

reviewed in the chapter by Hudgens[157] in this volume.

The recent studies performed at Yale University[158-160] on the CH radical provide a good example of the power of REMPI-PES in the study of transient species. The CH radicals were prepared by photodissociation of ketene or t-butylnitrite, and the (2+1) ionization spectrum was recorded between 285 and 330 nm by monitoring the m/e = 13 ion signal.[158-160] As in a number of other studies, a single laser was used both to photodissociate the precursor and to probe the CH fragment. In earlier flash photolysis spectra,[161] the bands in this region were obscured by strong absorption band of the diazomethane that was used to generate the CH radicals. The REMPI technique removes this difficulty and the (2+1) ionization spectra reveal new bands and allow the clarification of previous assignments.[158-160]

Because the ionic states of CH are quite well known from conventional spectroscopy, the REMPI-PES spectra are not expected to reveal new information on the energy levels of CH^+.[35] However, the knowledge of the ionic states and the REMPI-PES spectra can be used together to aid in the assignment of the (2+1) REMPI spectra. In the work on CH, REMPI-PES spectra[159] allowed the assignment of vibrational quantum numbers to spectral features in congested regions. The idea behind this can be understood as follows. As was discussed above, the photoionization of a Rydberg state is expected to occur without a change in vibrational quantum number. If the ionic energy levels are known. one can assign a particular vibrational quantum number, v', to Rydberg states that give strong $v^+=v'$ photoelectron peaks. Perturbations will produce photoelectron spectra with more than one strong peak. Although valence states produce more complicated photoelectron spectra, in the absence of perturbations these spectra will be the same for all rotational levels. Pallix et al.[159] have recorded constant ionic state (CIS) spectra in a particularly complicated region of the (2+1) ionization of CH. These spectra were recorded by monitoring a particular photoelectron peak (corresponding to the production of a particular ionic state) and scanning the wavelength.

Comparison of CIS spectra for the ionic states of interest clearly showed which features in the CH spectra have particular vibrational quantum numbers.[158-160] The use of REMPI-PES to assign REMPI spectra has been used in other studies, and is a potentially powerful application of the technique.

5. PHOTOELECTRON SPECTROSCOPY OF LARGE MOLECULES
5.1 Determination of Ionic Energy Levels

For the past twenty years, single photon UV-PES has been perhaps the best source of information on the vibrational structure of large molecular ions. The use of REMPI-PES offers a number of advantages over UV-PES. These have been summarized by Long et al.,[21] and include the improved linewidth of the light source, which leads to improved energy resolution, the ability to selectively excite a narrow rotational distribution, which reduces the rotational contour of the photoelectron band, and the reduced Doppler broadening achieved by the use of a supersonic molecular beam.[21] The most important advantage of REMPI-PES is the ability to record photoelectron spectra from a large number of resonant intermediate levels, effectively allowing one to choose the geometry of the ionized level in order to enhance the ionic vibrational modes of interest.

A detailed knowledge of the normal modes of the ground state and the excited neutral state greatly aids in the assignment of the ionic vibrational structure observed in the REMPI-PES spectra.[21,162] In the ideal case, the normal modes of the ground state and resonant intermediate state would all be known, and the geometry of the resonant intermediate state would be nearly the same as that of the ion. In this case, the $\Delta v=0$ propensity rule[47] for ionization would be expected to hold quite well, and the known assignments of the intermediate state would allow a straightforward assignment of the ionic spectrum. In addition, by exciting hot-band transitions to the same vibronic intermediate states, the consistency of the assignments could be checked.

In practice, the situation is not so ideal and the analysis must sometimes be made in an iterative fashion, with assignments of the vibrational structure of the ion leading to refinements of the assignments in the resonant intermediate state, and so forth. In addition, theoretical analysis is also extremely useful in clarifying assignments. However, although the analysis of REMPI-PES spectra of large molecules is non-trivial, studies on aniline,[38] benzene,[21] chlorobenzene,[162,163] and phenol[164] have greatly improved our understanding and knowledge of the vibrational structure of these ions. For example, Long et al.[21] have performed a detailed study of Jahn-Teller effects in $C_6H_6^+$ and $C_6D_6^+$ and have made the first observation of the Jahn-Teller splittings in these important ions. As another example, Anderson et al.[164] have performed a detailed study of the vibrational structure of the phenol ion, which allows the derivation of a harmonic force field for the in-plane vibrational modes. When combined with similar calculations for the neutral ground state and resonant intermediate state, this provides detailed information on the changes in bonding that occur with the removal of an electron from the molecular orbital system.[164]

The threshold photoelectron spectrometer described by Müller-Dethlefs et al.[25] has also been used to provide information on large molecular ions. In particular, Chewter et al. have recently published zero-kinetic-energy photoelectron spectroscopic determinations of the ionization potentials of benzene[26] and the benzene-argon van der Waals molecule.[27] It was also demonstrated that this technique can be used to observe the rotational structure of the benzene ion,[26] which would be extremely difficult to even imagine using dispersive electron energy analyzers.

5.2 REMPI-PES as a Probe of Intramolecular Relaxation Processes

Studies of large molecules using REMPI-PES may also be helpful in understanding dynamical processes occurring in the resonant intermediate state. For example, Achiba et al.[63,165] have performed two REMPI-PES studies on the intramolecular relaxation processes in

excited states of benzene, and the same group has also published a
REMPI-PES study on the intramolecular relaxation processes in
napthalene.[166] As an example, above a particular total excitation
energy, the REMPI-PES spectra through different vibronic modes of the
S_1 electronic state of napthalene show broadening in the observed
photoelectron bands.[166] This can be explained in terms of
intramolecular vibrational redistribution of the initially prepared
vibronic level, which populates a large number of vibrational levels
within the S_1 manifold. Assuming that $\Delta v=0$ ionizing transitions are
dominant, the photoelectron band for a given vibronic mode will be
shifted more or less as determined by the vibrational frequency
differences between the S_1 state and the ionic state. Thus, the
overlap of photoelectron bands resulting from $\Delta v=0$ ionization of the
large number of vibronic levels populated by IVR will lead to a
broadening. This type of broadening has been cited as the source of
broadening in the napthalene spectra.[166]

The REMPI-PES spectrum via the S_2 origin band of napthalene,
which occurs at somewhat higher energy, was also recorded.[166] If
ionization of the S_2 level was fast enough, one would expect to see a
strong photoelectron peak corresponding to the production of
vibrationless napthalene ions. Instead, a broad distribution of
photoelectron bands is observed, with no signal at the energy expected
for the vibrationless ion. This is interpreted[166] in terms of a fast
radiationless transition from the S_2 level, leading to highly
vibrationally excited molecules in the S_1 manifold. The photoelectron
spectrum shows significant intensity for vibrationally excited ions,
which is consistent with this explanation. The results indicate that
the radiationless transition rate must be high enough to compete with
the ionization rate of the initially prepared level. Hiraya et al.[166]
estimate the ionization rate for nanosecond lasers to be about 10^{11}
sec^{-1}, thus the relaxation rate must be of at least this order.

Pallix and Colson[18] have performed a similar study on the REMPI-
PES of sym-Triazine via the S_1 state. In these experiments,
photoelectron spectra were recorded at several wavelengths using 5

psec pulses and again using 2 nsec pulses to obtain information on the decay rate of the S_1 level. The photoelectron spectra obtained using picosecond pulses all show two strong bands corresponding to ionization of the prepared S_1 state as well as ionization of the T_1 state populated by intersystem crossing. The spectra obtained using nanosecond pulses also show two bands, however, they display substantially different intensities.[18] In some cases the spectra indicate that there is little population remaining in the S_1 state, while another spectrum displays an S_1 band that is nearly as intense as the T_1 band. A more detailed comparison of the spectra allows the conclusions that the coupling between the S_1 and T_1 manifolds is very inhomogeneous, and that the intersystem crossing rates vary from 10^7 sec^{-1} to $\gtrsim 10^{11}$ sec^{-1}.[18]

An intriguing extension of this picosecond study would be to prepare the resonant intermediate level and subsequently ionize it with a delayed probe. By choosing a system with the appropriate range of decay rates, and by varying the time delay of the probe pulse, it should be possible to monitor the time evolution of the initially prepared state. With sufficient knowledge of the resonant intermediate state and with sufficient photoelectron energy resolution, it may even be possible to determine which modes the vibrational energy flows into first, providing detailed information on the dynamics of radiationless transitions. Studies of this nature are being planned by research groups at a number of institutions.

6. SUMMARY

From the examples of the present chapter, it is clear that REMPI-PES is a valuable technique for probing excited state photoionization dynamics, for obtaining photoelectron spectra of transient species, for probing the vibrational structure of larger molecular ions, and for probing the intramolecular dynamics of excited neutral states. The impact of REMPI-PES techniques has just begun to be felt, and there will be enormous potential for growth in all of these areas for many years to come.

References

1. K. Siegbahn, C. Nordling, G. Johansson, J. Hedman, P. F. Heden, K. Hamrin, U. Gelius, T. Bergmark, L. O. Werme, R. Manne, and Y. Baer, ESCA Applied to Free Molecules (North-Holland, Amsterdam, 1969).

2. D. W. Turner, A. D. Baker, C. Baker, and C. R. Brundle, Molecular Photoelectron Spectroscopy, A Handbook of He 584Å Spectra (Interscience, New York, 1970).

3. J. H. D. Eland, Photoelectron Spectroscopy (Wiley-Halsted, New York, 1974).

4. J. W. Rabelais, Principles of Ultraviolet Photoelectron Spectroscopy (Wiley-Interscience, New York, 1977).

5. J. Berkowitz, Photoabsorption, Photoionization, and Photoelectron Spectroscopy (Academic, New York, 1979).

6. K. Kimura, S. Katsumata, Y. Achiba, T. Yamazaki, and S. Iwata, Handbook of HeI Photoelectron Spectra of Fundamental Organic Molecules (Halsted, New York, 1981).

7. See, for example, J. L. Dehmer, D. Dill, and A. C. Parr, in Photophysics and Photochemistry in the Vacuum Ultraviolet, eds. S. McGlynn, G. Findley, and R. Huebner (D. Reidel, Dordrecht, Holland, 1985), p. 341.

8. K. Kimura, Adv. Chem. Phys. $\underline{60}$, 161 (1985); K. Kimura, Int. Rev. Phys. Chem. (to be published).

9. J. P. Reilly, Israel J. Chem. $\underline{24}$, 266 (1984).

10. P. M. Dehmer, J. L. Dehmer, and S. T. Pratt, Comments At. Mol. Phys. $\underline{19}$, 205 (1987).

11. J. M. Dyke, N. Jonathan, and A. Morris, in Electron Spectroscopy: Theory, Techniques, and Applications, Vol. 3, edited by C. R. Brundle and A. D. Baker (Academic, New York, 1979) p. 189.

12. J. M. Dyke, N. Jonathan, and A. Morris, Int. Rev. Phys. Chem. $\underline{2}$, 3 (1982).

13. H. van Lonkhuyzen and C. A. DeLange, Mol. Phys. $\underline{51}$, 551 (1984).

14. P. M. Dehmer and J. L. Dehmer, J. Chem. Phys. $\underline{67}$, 1774 (1977).

15. P. M. Dehmer and J. L. Dehmer, J. Chem. Phys. $\underline{68}$, 3462 (1978).

16. P. M. Dehmer and J. L. Dehmer, J. Chem. Phys. 69, 125 (1978).

17. S. Tomoda and K. Kimura, Bull. Chem. Soc. Jpn. 56, 1768 (1983).

18. J. B. Pallix and S. D. Colson, Chem. Phys. Lett. 119, 38 (1985).

19. J. T. Meek, S. R. Long, and J. P. Reilly, J. Phys. Chem. 86, 2809 (1982).

20. J. Kimman, P. Kruit, and M. J. van der Wiel, Chem. Phys. Lett. 88, 576 (1982).

21. S. R. Long, J. T. Meek, and J. P. Reilly, J. Chem. Phys. 79, 3206 (1983).

22. P. Lambropoulos, Adv. At. Mol. Phys. 12, 87 (1976); S. N. Dixit and P. Lambropoulos, Phys. Rev. A 27, 861 (1983).

23. P. Kruit and F. Read, J. Phys. E 16, 313 (1983).

24. B. G. Koenders, K. E. Drabe, and C. A. DeLange, Chem. Phys. Lett. 138, 1 (1987).

25. K. Müller-Dethlefs, M. Sander, and E. W. Schlag, Z. Naturforsch. 39a, 1089 (1984); K. Müller-Dethlefs, M. Sander, and E. W. Schlag, Chem. Phys. Lett. 112, 291 (1984).

26. L. A. Chewter, M. Sander, K. Müller-Dethlefs, and E. W. Schlag, J. Chem. Phys. 86, 4737 (1987).

27. L. A. Chewter, K. Müller-Dethlefs, and E. W. Schlag, Chem. Phys. Lett. 135, 219 (1987).

28. S. T. Pratt, P. M. Dehmer, and J. L. Dehmer, J. Chem. Phys. 78, 4315 (1983).

29. S. T. Pratt, P. M. Dehmer, and J. L. Dehmer, J. Chem. Phys. 85, 3379 (1986).

30. M. A. O'Halloran, S. T. Pratt, P. M. Dehmer, and J. L. Dehmer, J. Chem. Phys. 87, 3288 (1987).

31. W. G. Wilson, K. S. Viswanathan, E. Sekreta, and J. P. Reilly, J. Phys. Chem. 88, 672 (1984).

32. K. S. Viswanathan, E. Sekreta, E. R. Davidson, and J. P. Reilly, J. Phys. Chem. 90, 5078 (1986).

33. R. L. Dubs, S. N. Dixit, and V. McKoy, J. Chem. Phys. 85, 656 (1986).

34. S. T. Pratt, P. M. Dehmer, and J. L. Dehmer, Chem. Phys. Lett. 105, 28 (1984).

35. K. P. Huber and G. Herzberg, Molecular Spectra and Molecular Structure IV. Constants of Diatomic Molecules (Van Nostrand Reinhold, New York, 1979).

36. I. Dabrowski and G. Herzberg, Can J. Phys. 66, 5584 (1977).

37. S. N. Dixit, D. L. Lynch, and V. McKoy, Phys. Rev. A 30, 3332 (1984); private communication.

38. S. N. Dixit and V. McKoy, J. Chem. Phys. 82, 3546 (1985).

39. S. N. Dixit, D. L. Lynch, V. McKoy, and W. M. Huo, Phys. Rev. A 32, 1267 (1985).

40. D. L. Lynch, S. N. Dixit, and V. McKoy, Chem. Phys. Lett. 123, 315 (1986).

41. S. N. Dixit and V. McKoy, Chem. Phys. Lett. 128, 49 (1986).

42. H. Rudolph, D. L. Lynch, S. N. Dixit, and V. McKoy, J. Chem. Phys. 84, 6657 (1986).

43. G. Herzberg, Spectra of Diatomic Molecules (Van Nostrand, Princeton, 1950).

44. D. Dill, Phys. Rev. A 6, 160 (1972).

45. D. Dill, in Photoionization and Other Probes of Many Electron Interactions, ed. F. Wuilleumier (Plenum, New York, 1976) p. 387.

46. U. Fano and D. Dill, Phys. Rev. A 6, 185 (1972).

47. R. S. Berry, J. Chem. Phys. 45, 1228 (1966).

48. W. A. Chupka, J. Chem. Phys. 87, 1488 (1987).

49. C. Cornaggia, A. Giusti-Suzor, and Ch. Jungen, J. Chem. Phys. 87, 3934 (1987).

50. A. P. Hickman, Phys. Rev. Lett. 59, 1553 (1987).

51. S. L. Guberman, J. Chem. Phys. 78, 1404 (1983).

52. H. J. M. Bonnie, P. J. Eenschuistra, P. J. Los, and H. J. Hopman, Chem. Phys. Lett. 125, 27 (1986).

53. H. J. M. Bonnie, J. W. J. Verschuur, H. J. Hopman, and H. B. van Linden van den Heuvell, Chem. Phys. Lett. 130, 43 (1986).

54. E. Xu, T. Tsuboi, R. Kachru, and H. Helm, Phys. Rev. A (in press).

55. I. Dabrowski, Can J. Phys. 62, 1639 (1984).

56. S. T. Pratt, P. M. Dehmer, and J. L. Dehmer, J. Chem. Phys. 86, 1727 (1987).

57. S. T. Pratt, P. M. Dehmer, and J. L. Dehmer, J. Chem. Phys. 87, 4423 (1987).

58. I. Dabrowski and G. Herzberg, Can. J. Phys. 52, 1110 (1974).

59. Y. Achiba, K. Sato, K. Shobatake, and K. Kimura, J. Chem. Phys. 78, 5474 (1983).

60. S. T. Pratt, P. M. Dehmer, and J. L. Dehmer, J. Chem. Phys. 80, 1706 (1984).

61. J. H. Glownia, S. J. Riley, S. D. Colson, J. C. Miller, and R. N. Compton, J. Chem. Phys. 77, 68 (1982).

62. W. E. Conaway, R. J. S. Morrison, and R. N. Zare, Chem. Phys. Lett. 113, 429 (1985).

63. Y. Achiba, K. Sato, K. Shobatake, and K. Kimura, J. Chem. Phys. 79, 5213 (1983).

64. J. T. Meek, E. Sekreta, W. Wilson, K. S. Viswanathan, and J. P. Reilly, J. Chem. Phys. 82, 1741 (1985).

65. D. Stahel, M. Leoni, and K. Dressler, J. Chem. Phys. 79, 2541 (1983).

66. A. Lofthus and P. H. Krupenie, J. Phys. Chem. Ref. Data 6, 113 (1977).

67. See, for example, J. Ganz, B. Lewandowski, A. Siegel, W. Bussert, H. Waibel, M. W. Ruf, and H. Hotop, J. Phys. B 15, L485 (1982); A. Siegel, J. Ganz, W. Bussert, and H. Hotop, J. Phys. B 16, 2945 (1983), and references therein.

68. W. E. Conaway, T. Ebata, and R. N. Zare, J. Chem. Phys. 87, 3447 (1987).

69. W. E. Conaway, T. Ebata, and R. N. Zare, J. Chem. Phys. 87, 3453 (1987).

70. M. G. White, M. Seaver, W. A. Chupka, and S. D. Colson, Phys. Rev. Lett. 49, 28 (1982).

71. S. T. Pratt, P. M. Dehmer, and J. L. Dehmer, J. Chem. Phys. 81, 3444 (1984).

72. H. H. Michels, Adv. Chem. Phys. 45, 225 (1981).

73. See, for example, H. Lefebvre-Brion, in Atoms, Molecules, and Lasers (International Atomic Energy Agency, Vienna, 1974) p. 411.

74. P. K. Carroll and C. P. Collins, Can J. Phys. 47, 563 (1969).

75. S. L. Anderson, G. D. Kubiak, and R. N. Zare, Chem. Phys. Lett. 105, 22 (1984).

76. M. G. White, W. A. Chupka, M. Seaver, A. Woodward, and S. D. Colson, J. Chem. Phys. 80, 678 (1984).

77. M. Leoni and K. Dressler, Z. Angew. Math. Phys. 22, 794 (1971).

78. K. Yoshino, Y. Tanaka, P. K. Carroll, and P. Mitchell, J. Mol. Spectrosc. 54, 87 (1975).

79. J. J. Hopfield, Nature (London) 125, 927 (1930).

80. O. W. Richardson, Molecular Hydrogen and Its Spectrum (Yale University Press, New Haven, CT, 1934).

81. (a) H. Beutler, Z. Phys. Chem. (Frankfurt) B 29, 315 (1935); (b) H. Beutler, A. Deubner, and H. O. Jünger, Z. Phys. 98, 181 (1936); (c) H. Beutler and H. O. Jünger, Z. Phys. 100, 80 (1936); (d) H. Beutler and H. O. Jünger, 101, 285, 304 (1936).

82. Y. Tanaka, Sci. Papers Inst. Phys. Chem. Res. (Tokyo) 42, 49 (1944).

83. A. Monfils, J. Mol. Spectrosc. 15, 265 (1965) and references therein.

84. (a) T. Namioka, J. Chem. Phys. 40, 3154 (1964); (b) T. Namioka, J. Chem. Phys. 41, 2141 (1964); and (c) T. Namioka, J. Chem. Phys. 43, 1636 (1965).

85. S. Takezawa, J. Chem. Phys. 52, 2575, 5793 (1970).

86. G. Herzberg and Ch. Jungen, J. Mol. Spectrosc. 41, 425 (1972).

87. G. R. Cook and P. H. Metzger, J. Opt. Soc. Am. 54, 968 (1964).

88. V. H. Dibeler, R. M. Reese, and M. Krauss, J. Chem. Phys. 42, 2045 (1965).

89. F. J. Comes and H. O. Wellern, Z. Naturforsch. Teil A 23, 881 (1968).

90. W. A. Chupka and J. Berkowitz, J. Chem. Phys. 51, 4244 (1969).

91. P. M. Dehmer and W. A. Chupka, J. Chem. Phys. 65, 2243 (1976).

92. W. L. Glab and J. P. Hessler, Phys. Rev. A 35, 2102 (1987).

93. N. Bjerre, R. Kachru, and H. Helm, Phys. Rev. A 31, 1206 (1985).

94. D. M. Parker, J. D. Buck, and D. W. Chandler, J. Phys. Chem. 91, 2035 (1987).

95. E. Y. Xu, H. Helm, and R. Kachru, Phys. Rev. Lett. 59, 1096 (1987).

96. (a) M. J. Seaton, Proc. Phys. Soc. London 88, 801 (1966); (b) M. J. Seaton, Mon. Not. R. Astron. Soc. 118, 504 (1958); for a recent review see M. J. Seaton, Rep. Prog. Phys. 46, 167 (1983).

97. U. Fano, Phys. Rev. A 2, 353 (1970).

98. D. Dill and Ch. Jungen, J. Phys. Chem. 84, 2116 (1980).

99. Ch. Jungen and D. Dill, J. Chem. Phys. 73, 3338 (1980).

100. Ch. Jungen, in Electronic and Atomic Collisions, edited by S. Datz (North-Holland, Amsterdam, 1982) p. 455.

101. Ch. Jungen, J. Chim. Phys. 77, 27 (1980).

102. Ch. Jungen, Phys. Rev. Lett. 53, 2394 (1984).

103. Ch. Jungen and M. Raoult, Faraday Discuss. Chem. Soc. 71, 253 (1981).

104. M. Raoult, Ch. Jungen, and D. Dill, J. Chim. Phys. 77, 599 (1980).

105. M. Raoult and Ch. Jungen, J. Chem. Phys. 74, 3388 (1981).

106. S. E. Nielsen and R. S. Berry, Chem. Phys. Lett. 2, 503 (1968).

107. R. S. Berry and S. E. Nielsen, Phys. Rev. A 1, 383, 395 (1970).

108. J. N. Bardsley, Chem. Phys. Lett. 1, 229 (1967).

109. G. B. Shaw and R. S. Berry, J. Chem. Phys. 56, 5808 (1972).

110. J. Berkowitz and W. A. Chupka, J. Chem. Phys. 51, 2341 (1969).

111. P. M. Dehmer and W. A. Chupka, J. Chem. Phys. 66, 1972 (1977).

112. E. E. Eyler, J. Gilligan, E. McCormack, A. Nussenzweig, and E. Pollack, Phys. Rev. A 36, 3486 (1987).

113. E. E. Marinero, C. T. Rettner, and R. N. Zare, Phys. Rev. Lett. 48, 1323 (1982).

114. E. E. Marinero, R. Vasudev, and R. N. Zare, J. Chem. Phys. 78, 692 (1983).

115. S. L. Anderson, G. D. Kubiak, and R. N. Zare, Chem. Phys. Lett. 105, 22 (1987).

116. J. Kimman, J. W. J. Verschuur, M. Lavollee, H. B. van Linden van den Heuvell, and M. J. van der Wiel, J. Phys. B 19, 3909 (1986).

117. J. W. J. Verschuur, J. Kimman, H. B. van Linden van den Heuvell, and M. J. van der Wiel, Chem. Phys. 103, 359 (1986).

156

118. Y. Achiba and K. Kimura, Book of Abstracts, International Conference on Multiphoton Processes III, Crete, 1984.

119. A. Giusti-Suzor and Ch. Jungen, J. Chem. Phys. 80, 989 (1984).

120. S. Fredin, D. Gauyacq, M. Horani, Ch. Jungen, and G. Lefevre, Mol. Phys. 60, 825 (1987).

121. S. T. Pratt, P. M. Dehmer, and J. L. Dehmer, J. Chem. Phys. 85, 5535 (1986).

122. G. Leuchs and H. Walther, in Multiphoton Ionization of Atoms, edited by S. L. Chin and P. Lambropoulos (Academic, New York, 1984) p. 109.

123. S. N. Dixit and P. Lambropoulos, Phys. Rev. Lett. 46, 1278 (1981); S. N. Dixit and P. Lambropoulos, Phys. Rev. A 27, 168 (1983).

124. Y. Ono, S. H. Linn, H. F. Prest, C. Y. Ng, and E. Miescher, J. Chem. Phys. 73, 4855 (1980).

125. M. E. Wacks, J. Chem. Phys. 41, 930 (1964); M. Halmann and I. Laulicht, J. Chem. Phys. 43, 1503 (1965).

126. C. N. Yang, Phys. Rev. 74, 764 (1948).

127. See, for example, C. H. Greene and R. N. Zare, Ann. Rev. Phys. Chem. 33, 119 (1982); C. H. Greene and R. N. Zare, J. Chem. Phys. 78, 6741 (1983).

128. W. J. Kessler and E. D. Poliakoff, J. Chem. Phys. 84, 3647 (1986).

129. R. L. Dubs, S. N. Dixit, and V. McKoy, Phys. Rev. Lett. 54, 1249 (1985).

130. R. L. Dubs, S. N. Dixit, and V. McKoy, J. Chem. Phys. 86, 5886 (1987).

131. R. L. Dubs, S. N. Dixit, and V. McKoy, J. Chem. Phys. 85, 6267 (1986).

132. J. R. Appling, M. G. White, T. M. Orlando, and S. L. Anderson, J. Chem. Phys. 85, 6803 (1986).

133. K. Sato, Y. Achiba, and K. Kimura, J. Chem. Phys. 81, 57 (1984).

134. K. Fuke, H. Yoshiuchi, K. Kaya, Y. Achiba, K. Sato, and K. Kimura, Chem. Phys. Lett. 108, 179 (1984).

135. S. T. Pratt, P. M. Dehmer, and J. L. Dehmer, Chem. Phys. Lett. 116, 245 (1985).

136. S. T. Pratt, P. M. Dehmer, and J. L. Dehmer, J. Chem. Phys. 82, 5758 (1985).

137. P. M. Dehmer, S. T. Pratt, and J. L. Dehmer, J. Phys. Chem. 91, 2593 (1987).

138. P. M. Dehmer and S. T. Pratt, J. Chem. Phys. (submitted).

139. E. D. Poliakoff, P. M. Dehmer, J. L. Dehmer, and R. Stockbauer, J. Chem. Phys. 75, 1568 (1981).

140. E. D. Poliakoff, P. M. Dehmer, J. L. Dehmer, and R. Stockbauer, J. Chem. Phys. 76, 5214 (1982).

141. C. E. Moore, Nat'l. Bur. Stand. Circ. 466 (1949).

142. See, for example, P. M. Johnson and C. E. Otis, Ann. Rev. Phys. Chem. 32, 139 (1981); M. N. R. Ashfold, Mol. Phys. 58, 1 (1986).

143. P. M. Dehmer and S. T. Pratt, J. Chem. Phys. 77, 4804 (1982).

144. M. V. Bobetic and J. A. Barker, J. Chem. Phys. 64, 2367 (1976); J. K. Lee, D. Henderson, and J. A. Barker, Mol. Phys. 29, 429 (1975).

145. S. R. Leone, Adv. Chem. Phys. 50, 255 (1982), and references therein.

146. P. M. Dehmer, S. T. Pratt, and J. L. Dehmer, J. Phys. Chem. 91, 2593 (1987).

147. R. S. Mulliken, J. Chem. Phys. 52, 5170 (1970).

148. R. S. Mulliken, Radiat. Res. 59, 357 (1974).

149. H. H. Michels, R. H. Hobbs, and L. A. Wright, J. Chem. Phys. 69, 5151 (1978).

150. W. R. Wadt, J. Chem. Phys. 68, 402 (1978).

151. P. M. Dehmer, S. T. Pratt, and J. L. Dehmer, J. Chem. Phys. 85, 13 (1986).

152. C. Y. Ng, D. J. Trevor, B. H. Mahan, and Y. T. Lee, J. Chem. Phys. 65, 4327 (1976).

153. S. T. Pratt, J. L. Dehmer, and P. M. Dehmer, J. Chem. Phys. 82, 676 (1985).

154. S. T. Pratt, J. L. Dehmer, and P. M. Dehmer, Phys. Rev. A 36, 1702 (1987).

155. C. A. DeLange, private communication. (1987).

156. S. T. Pratt, Phys. Rev. A 33, 1718 (1986).

157. J. W. Hudgens, in this volume.

158. P. Chen, W. A. Chupka, and S. D. Colson, Chem. Phys. Lett. 121, 405 (1985).

159. J. B. Pallix, P. Chen, W. A. Chupka, and S. D. Colson, J. Chem. Phys. 84, 5208 (1986).

160. P. Chen, J. B. Pallix, W. A. Chupka, and S. D. Colson, J. Chem. Phys. 86, 516 (1987).

161. G. Herzberg and J. W. C. Johns, Astrophys, J. 158, 399 (1969).

162. S. L. Anderson, D. M. Rider, and R. N. Zare, Chem. Phys. Lett. 93, 11 (1982).

163. J. L. Durant, D. M. Rider, S. L. Anderson, F. D. Proch, and R. N. Zare, J. Chem. Phys. 80, 1817 (1984).

164. S. L. Anderson, L. Goodman, K. Krogh-Jesperson, A. G. Ozkabak, R. N. Zare, and C. Zheng, J. Chem. Phys. 82, 5329 (1985).

165. Y. Achiba, A. Hiraya, and K. Kimura, J. Chem. Phys. 80, 6047 (1984).

166. A. Hiraya, Y. Achiba, N. Mikami, and K. Kimura, J. Chem. Phys. 82, 1810 (1985).

APPENDIX

SUMMARY OF MOLECULAR REMPI-PES STUDIES

Molecule	Excited State	Process	Reference
H_2	B $^1\Sigma_u^+$, v=7	(3+1)	A13
H_2	B $^1\Sigma_u^+$, v=8-11	(3+1)	A47
H_2	C $^1\Pi_u$, v=0-4	(3+1)	A19
H_2	C $^1\Pi_u$, v=0-4	(3+1)	A52
H_2	C $^1\Pi_u$, v=0-4	(3+1)	A54
H_2	C $^1\Pi_u$, v=0-4	(3+1)	A64
H_2	B' $^1\Sigma_u^+$, v=0-2	(3+1)	A53
	D $^1\Pi_u$, v=0,1	(3+1)	
H_2	E,F $^1\Sigma_g^+$, v_E=0-2	(2+1)	A30
H_2	E,F $^1\Sigma_g^+$, v_E=1-3	(4+1)	A48
		(4+2)ATI	
H_2	E,F $^1\Sigma_g^+$, v_E=0	(4+2)ATI	A61
D_2	C $^1\Pi_u$, v=0-4	(3+1)	A77
N_2	o_3 $^1\Pi_u$, v=1,2	(3+1)	A20
N_2	b $^1\Pi_u$, v=0-5	(3+1)	A21
	c $^1\Pi_u$, v=0,1	(3+1)	
	c' $^1\Sigma_u^+$, v=0,1	(3+1)	
CO	A $^1\Pi$, v=1,2	(3+3)	A12
	A $^1\Pi$, v=3	(3+2)	
CO	A $^1\Pi$, v=1-3	(2+2)	A14
NO	A $^2\Sigma^+$, v=0-3	(2+2)	A2
	C $^2\Pi$, v=0,1	(2+1)	
NO	A $^2\Sigma^+$, v=0,1	(2+2)	A10
NO	C $^2\Pi$, v=0	(3+2)	A5
NO	F $^2\Delta$, v=0,1	(3+1)	A15
	H $^2\Sigma^+$,H' $^2\Pi$, v=0,1	(3+1)	
NO	A $^2\Sigma^+$, v=0	(1+1)	A27
NO	C $^2\Pi$, v=0	(3+1)	A29
NO	C $^2\Pi$, v=0	(3+1)	A28

SUMMARY OF MOLECULAR REMPI-PES STUDIES (Continued)

Molecule	Excited State	Process	Reference
NO	A $^2\Sigma^+$, v=0	(2+2)	A23
NO	C $^2\Pi$	(2+1)	A39, A65
	B $^2\Pi$	(2+1)	
	L $^2\Pi$	(2+1)	
NO	B $^2\Pi$, v=9	(2+1)	A37
NO	A $^2\Sigma^+$, v=0,1	(2+2)	A45
	A $^2\Sigma^+$, v=0	(1+1)	
	E $^2\Sigma^+$, v=0,1	(2+1')	
	H $^2\Sigma^+$, H' $^2\Pi$, v=1	(2+1')	
	F $^2\Delta$, v=3	(1+1)	
	N $^2\Delta$, v=1	(1+1)	
NO	A $^2\Sigma^+$, v=1	(2+1')	A46
NO	Autoionizing Rydberg states	(2+0)	A51
NO	A $^2\Sigma^+$, v=0,1	(2+2) (2+1')	A7
NO	A $^2\Sigma^+$, v=0	(1+1)	A60
NO	High Rydberg states, v=3	(2+1')	A66
NO	C $^2\Pi$, v=0	(2+1)	A74
	D $^2\Sigma^+$, v=0	(2+1)	
	A $^2\Sigma^+$, v=0	(1+1)	
NO	A $^2\Sigma^+$, v=0, CDAD	(2+1')	A71
(NO)$_2$	Near NO A $^2\Sigma^+$	(1+1)	A62
O$_2$	C $^3\Pi_g$, v=2	(2+1)	A55, A56
CH	D $^2\Pi$, v=2	(2+1)	A43
Cl$_2$	2Π_u and $2^1\Sigma_u^+$	(3+1) (3+2)	A79
Br$_2$	Rydberg states	(3+1)	A63
CF	B $^2\Delta$, v=2	(1+1)	A69

SUMMARY OF MOLECULAR REMPI-PES STUDIES (Continued)

Molecule	Excited State	Process	Reference
CCl		(2+0)	A69
H_2S	4A-1, (1,0,0)	(3+1)	A6
H_2S	$3A_2$, 3C, 3D	(3+1)	A8
	4A-1, 4A-2, 4A-3,		
	4A-4, $4A_1$, $4A_2'$		
NH_3	B $^1E''$, v=3-10	(2+1)	A40
	C' $^1A_1'$, v=0-6	(2+1)	
NH_3	C' $^1A_1'$, v=0-9	(2+1)	A70,A72,A73
NH_3	B $^1E''$, v=7-11	(3+1)	A4
	C' $^1A_1'$, v=2-5	(3+1)	
	D $^1E''$, v=0-2,8,9	(3+1)	
	A' $^1A_2''$, v=1,2	(2+2)	
NH_3	B $^1E''$, v=9-11	(2+1)	A44
	C' $^1A_1'$, v=5-7	(2+1)	
	D $^1E''$, v=0,1	(2+1)	
NH_3	C' $^1A_1'$, v=0-5	(3+1)	A15
CH_3I	$^1\Sigma^+$, v=0	(2+1)	A42
	$^1\Pi$, v=0	(2+1)	
CH_3I	Σ, v=0,1	(2+1)	A49
	Π, v=0,1	(2+1)	
ArXe	Rydberg diss.	(2+1)	A33
	to Ar+Xe* $6p[1/2]_0$		
KrXe	Rydberg diss.	(2+1)	A32
	to Kr+Xe* $6p[5/2]_2$		
Kr_2	Rydberg diss.	(2+1)	A78
	to Kr+Kr*$5p[1/2]_0$		
Xe_2	Rydbergs diss.	(2+1)	A50
	to Xe+Xe* 5d,6p		
ArNO	Rydberg diss.	(2+1)	A25
	to Ar+NO* C $^2\Pi$		

162

SUMMARY OF MOLECULAR REMPI-PES STUDIES (Continued)

Molecule	Excited State	Process	Reference
Acetylene	Ungerade Rydberg states between 74800 and 90000 cm^{-1}	(3+1)	A57
Acetylene	Gerade Rydberg states between 72400 and 77200 cm^{-1}	(2+1)	A58
1,3-trans-butadiene	B	(2+1)	A22
Benzene	A $^1B_{2u}$	(2+2)	A3
Benzene	"Johnson Band"	(2+1)	A1
Benzene	A $^1B_{2u}$	(1+1)	A17
Benzene	A $^1B_{2u}$	(2+2)	A16
	$^1E_{1u}$	(2+1+1)	
Benzene	A $^1B_{2u}$	(1+1)	A24
Benzene	S_2 $^1B_{1u}$	(1+1)	A59
Benzene	ZEKE-PES	(1+1')	A67
Ar-Benzene	ZEKE-PES	(1+1')	A68
Chlorobenzene	$^1B_2$	(1+1)	A11
Chlorobenzene	$^1B_2$	(1+1)	A31
Toluene	S_1	(1+1)	A9
Benzaldehyde	S_2	(1+1)	A18
Phenol	$^1B_2$	(1+1)	A41
Aniline	$^1B_2$	(1+1)	A38
Napthalene	S_1, S_2	(1+1)	A35
		(1+1')	
Trimethylamine	S_2	(1+1)	A75
Triethylamine	S_2	(2+1)	A36
s-Triazine	S_1 $^1E"$	(1+2)	A34
Phenol and Phenol-H_2O complexes	S_1	(1+1)	A26

SUMMARY OF MOLECULAR REMPI-PES STUDIES (Continued)

Molecule	Excited State	Process	Reference
7-Azaindole and 7-Azaindole-H_2O complexes	S_1	(1+1)	A26

164

APPENDIX REFERENCES

A1. J. C. Miller and R. N. Compton, J. Chem. Phys. 75, 2020 (1981),
[Benzene, (2+1) via the "Johnson Band" at 391.4 nm].

A2. J. C. Miller and R. N. Compton, J. Chem. Phys. 75, 22 (1981),
[NO, (2+2) via A $^2\Sigma^+$, v=0-3, and (2+1) via C $^2\Pi$, v=0,1].

A3. Y. Achiba, K. Sato, K. Shobatake, and K. Kimura, J. Photochem.
17, 199 (1981), [Benzene, (2+2) via A $^1B_{2u}$].

A4. J. H. Glownia S. J. Riley, S. D. Colson, J. C. Miller, and R. N.
Compton, J. Chem. Phys. 77, 68 (1982), [NH_3, (3+1) via B $^1E''$,
v=7-11; (3+1) via C' $^1A_1'$, v=2-5; (3+1) via D $^1E''$, v=0-2,8,9; and
(2+2) via A' $^1A_2''$, v=1,2].

A5. M. G. White, M. Seaver, W. A. Chupka, and S. D. Colson, Phys.
Rev. Lett. 49, 28 (1982), [NO, (3+2) via C $^2\Pi$, v=0].

A6. J. C. Miller, R. N. Compton, T. E. Carney, and T. Baer, J. Chem.
Phys. 76, 5648 (1982), [H_2S, (3+1) via 4A-1].

A7. J. C. Miller and R. N. Compton, Chem. Phys. Lett. 93, 453 (1982),
[NO, (2+2) and (2+1') via A $^2\Sigma^+$, v=0,1].

A8. Y. Achiba, K. Sato, K. Shobatake, and K. Kimura, J. Chem. Phys.
77, 2709 (1982), [H_2S, (3+1) via 3A$_2$, 3C, 3D, 4A-1, 4A-2, 4A-3,
4A-4, 4A$_1$, 4A$_2'$].

A9. J. T. Meek, S. R. Long, and J. P. Reilly, J. Phys. Chem. 86, 2809
(1982), [Toluene, (1+1) via S$_1$].

A10. J. Kimman, P. Kruit, and M. J. van der Wiel, Chem. Phys. Lett.
88, 576 (1982), [NO, (2+2) via A $^2\Sigma^+$, v=0,1].

A11. S. L. Anderson, D. M. Rider, and R. N. Zare, Chem. Phys. Lett.
93, 11 (1982), [Chlorobenzene, (1+1) via $^1B_2$].

A12. S. T. Pratt, E. D. Poliakoff, P. M. Dehmer, and J. L. Dehmer, J.
Chem. Phys. 78, 65 (1983), [CO, (3+3) via A $^1\Pi$, v=1,2; (3+2) via
A $^1\Pi$, v=3].

A13. S. T. Pratt, P. M. Dehmer, and J. L. Dehmer, J. Chem. Phys. 78,
4315 (1983), [H_2, (3+1) via B $^1\Sigma_u^+$, v=7].

A14. S. T. Pratt, P. M. Dehmer, and J. L. Dehmer, J. Chem. Phys. 79,
3234 (1983), [CO, (2+2) via A $^1\Pi$, v=1-3].

A15. Y. Achiba, K. Sato, K. Shobatake, and K. Kimura, J. Chem.
Phys. 78, 5474 (1983), [NO, (3+1) via F $^2\Delta$, v=0,1; (3+1) via
H $^2\Sigma^+$,H' $^2\Pi$, v=0,1; NH$_3$, (3+1) via C' $^1A_1'$, v=0-5].

A16. Y. Achiba, K. Sato, K. Shobatake, and K. Kimura, J. Chem. Phys.
79, 5213 (1983), [Benzene, (2+2) via A $^1B_{2u}$; (2+1+1) via $^1E_{1u}$].

A17. S. R. Long, J. T. Meek, and J. P. Reilly, J. Chem. Phys. 79, 3206
(1983), [Benzene, (1+1) via A $^1B_{2u}$].

A18. S. R. Long, J. T. Meek, P. J. Harrington, and J. P. Reilly, J.
Chem. Phys. 78, 3341 (1983), [Benzaldehyde, (1+1) via S$_2$].

A19. S. T. Pratt, P. M. Dehmer, and J. L. Dehmer, Chem. Phys. Lett.
105, 28 (1984), [H$_2$, (3+1) via C $^1\Pi_u$, v=0-4].

A20. S. T. Pratt, P. M. Dehmer, and J. L. Dehmer, J. Chem. Phys. 80,
1706 (1984), [N$_2$, (3+1) via o$_3$ $^1\Pi_u$, v=1,2].

A21. S. T. Pratt, P. M. Dehmer, and J. L. Dehmer, J. Chem. Phys. 81,
3444 (1984), [N$_2$, (3+1) via b $^1\Pi_u$, v=0-5; (3+1) via c $^1\Pi_u$, v=0-1;
(3+1) via c' $^1\Sigma_u^+$, v=0,1].

A22. A. M. Woodward, W. A. Chupka, and S. D. Colson, J. Phys. Chem.
88, 4567 (1984), [1,3-Trans-butadiene, (2+1) via B].

A23. M. G. White, W. A. Chupka, M. Seaver, A. Woodward, S. D. Colson,
J. Chem. Phys. 80, 678 (1984), [NO, (2+2) via A $^2\Sigma^+$, v=0].

A24. Y. Achiba, A. Hiraya, and K. Kimura, J. Chem. Phys. 80, 6047
(1984), [Benzene, (1+1) via A $^1B_{2u}$].

A25. K. Sato, Y. Achiba, and K. Kimura, J. Chem. Phys. 81, 57 (1984),
[ArNO, (2+1) via a Rydberg dissociating to Ar+NO* C $^2\Pi$].

A26. K. Fuke, H. Yoshiuchi, K. Kaya, Y. Achiba, K. Sato, and K.
Kimura, Chem. Phys. Lett. 108, 179 (1984), [Phenol, Phenol
clusters, Phenol-H$_2$O clusters, 7-Azaindole, 7-Azaindole clusters,
and 7-Azaindole-H$_2$O clusters, (1+1) via S$_1$].

A27. W. G. Wilson, K. S. Viswanathan, E. Sekreta, and J. P. Reilly, J.
Phys. Chem. 88, 672 (1984), [NO, (1+1) via A $^2\Sigma^+$, v=0].

A28. K. Müller-Dethlefs, M. Sander and E. W. Schlag, Z. Naturforsch.
39a, 1089 (1984), [NO, (3+1) via C $^2\Pi$, v=0].

A29. K. Müller-Dethlefs, M. Sander and E. W. Schlag, Chem. Phys. Lett.
112, 291 (1984), [NO, (3+1) via C $^2\Pi$, v=0].

A30. S. L. Anderson, G. D. Kubiak, and R. N. Zare, Chem. Phys. Lett. <u>105</u>, 22 (1984), [H_2, (2+1) E,F $^1\Sigma_g^+$, v_E=0-2].

A31. J. L. Durant, D. M. Rider, S. L. Anderson, F. D. Proch, and R. N. Zare, J. Chem. Phys. <u>80</u>, 1817 (1984), [Chlorobenzene, (1+1) via $^1B_2$].

A32. S. T. Pratt, P. M. Dehmer, and J. L. Dehmer, Chem. Phys. Lett. <u>116</u>, 245 (1985), [KrXe, (2+1) via a Rydberg state dissociating to Kr+Xe* 6p[5/2]$_2$].

A33. S. T. Pratt, P. M. Dehmer, and J. L. Dehmer, J. Chem. Phys. <u>82</u>, 5758 (1985), [ArXe, (2+1) via a Rydberg state dissociating to Ar+Xe* 6p[1/2]$_0$].

A34. J. B. Pallix and S. D. Colson, Chem. Phys. Lett. <u>119</u>, 38 (1985), [Sym-triazine, (1+2) via S_1 ^{1}E"].

A35. A. Hiraya, Y. Achiba, N. Mikami, and K. Kimura, J. Chem. Phys. <u>82</u>, 1810 (1985), [Napthalene, (1+1) and (1+1') via S_1 and S_2].

A36. M. Kawasaki, K. Kasatani, H. Sato, Y. Achiba, K. Sato, and K. Kimura, Chem. Phys. Lett. <u>114</u>, 473 (1985), [Triethylamine, (2+1) via S_2].

A37. Y. Achiba, K. Sato, and K. Kimura, J. Chem. Phys. <u>82</u>, 3959 (1985), [NO, (2+1) via B $^2\Pi$, v=9].

A38. J. T. Meek, E. Sekreta, W. Wilson, K. S. Viswanathan, and J. P. Reilly, J. Chem. Phys. <u>82</u>, 1741 (1985), [Aniline, (1+1) via $^1B_2$].

A39. J. Kimman, M. Lavollee, and M. J. van der Wiel, Chem. Phys. <u>97</u>, 137 (1985), [NO, (2+1) via C $^2\Pi$, B $^2\Pi$, and L $^2\Pi$].

A40. W. E. Conaway, R. J. S. Morrison, and R. N. Zare, Chem. Phys. Lett. <u>113</u>, 429 (1985), [NH_3, (2+1) NH_3 via B ^{1}E", v=3-10; (2+1) via C' $^1A_1'$, v=0-6]

A41. S. L. Anderson, L. Goodman, K. Krogh-Jespersen, A. G. Ozkabak, R. N. Zare, and C. Zheng, J. Chem. Phys. <u>82</u>, 5329 (1985), [Phenol, (1+1) via $^1B_2$].

A42. W. A. Chupka, A. M. Woodward, S. D. Colson, and M. G. White, J. Chem. Phys. <u>82</u>, 4880 (1985), [CH_3I, (2+1) via $^1\Sigma^+$, v=0; (2+1) via $^1\Pi$, v=0].

A43. J. B. Pallix, P. Chen, W. A. Chupka, and S. D. Colson, J. Chem. Phys. <u>84</u>, 5208 (1986), [CH, (2+1) via D $^2\Pi$, v=2].

A44. J. B. Pallix and S. D. Colson, J. Phys. Chem. $\underline{90}$, 1499 (1986), [NH_3, (2+1) via B $^1E''$, v=9-11; (2+1) via C' $^1A_1'$, v=5-7; (2+1) via D $^1E''$, v=0,1].

A45. J. C. Miller and R. N. Compton, J. Chem. Phys. $\underline{84}$, 675 (1986), [NO, (2+2) via A $^2\Sigma^+$, v=0,1; (1+1) via A $^2\Sigma^+$, v=0; (2+1') via E $^2\Sigma^+$, v=0,1; (2+1') via H $^2\Sigma^+$,H' $^2\Pi$, v=1; (1+1) via F $^2\Delta$, v=3; (1+1) via N $^2\Delta$, v=1].

A46. J. W. J. Verschuur, J. Kimman, H. B. van Linden van den Heuvell, and M. J. van der Wiel, Chem. Phys. $\underline{103}$, 359 (1986), [NO, (2+1') via A $^2\Sigma^+$, v=1].

A47. J. H. M. Bonnie, J. W. J. Verschuur, H. J. Hopman, and H. B. van Linden van den Heuvell, Chem. Phys. Lett. $\underline{130}$, 43 (1986), [H_2, (3+1) via B $^1\Sigma_u^+$, v=8-11].

A48. C. Cornaggia, D. Normand, J. Morellec, G. Mainfray, and C. Manus, Phys. Rev. A. $\underline{34}$, 207 (1986), [H_2, (4+1) and (4+2) above threshold ionization via E,F $^1\Sigma_g^+$, v_E = 1-3].

A49. A. M. Woodward, S. D. Colson, W. A. Chupka, and M. G. White, J. Phys. Chem. $\underline{86}$, (1986), [CH_3I, (2+1) via $^1\Sigma^+$, v=0,1; (2+1) via $^1\Pi$ v=0,1].

A50. P. M. Dehmer, S. T. Pratt, and J. L. Dehmer, J. Phys. Chem. $\underline{91}$, 2593 (1987), [Xe_2, (2+1) via Rydberg states dissociating to Xe+Xe* 5d,6p].

A51. S. T. Pratt, P. M. Dehmer, and J. L. Dehmer, J. Chem. Phys. $\underline{85}$, 5535 (1986), [NO, (2+0) via autoionizing Rydberg states].

A52. S. T. Pratt, P. M. Dehmer, and J. L. Dehmer, J. Chem. Phys. $\underline{85}$, 3379 (1986), [H_2, (3+1) via C $^1\Pi_u$, v=0-4].

A53. S. T. Pratt, P. M. Dehmer, and J. L. Dehmer, J. Chem. Phys. $\underline{86}$, 1727 (1987), [H_2, (3+1) via B' $^1\Sigma_u^+$, v=0-2; (3+1) via D $^1\Pi_u$, v=0,1].

A54. M. A. O'Halloran, J. L. Dehmer, S. T. Pratt, and P. M. Dehmer, J. Chem. Phys. $\underline{87}$, 3288 (1987), [H_2, (3+1) via C $^1\Pi_u$, v=0-4 and (3+1) via B $^1\Sigma_u^+$].

A55. A. Sur, C. V. Ramana, W. A. Chupka, and S. D. Colson, J. Chem. Phys. $\underline{84}$, 69 (1986), [O_2, (2+1) via C $^3\Pi_g$, v=2].

A56. S. Katsumata, K. Sato, Y. Achiba, and K. Kimura, J. Electron Spectrosc. Rel. Phen. $\underline{41}$, 325 (1986), [O_2, (2+1) via C $^3\Pi_g$, v=2].

A57. T. M. Orlando, S. L. Anderson, J. R. Appling, and M. G. White, J. Chem. Phys. $\underline{87}$, 852 (1987), [C_2H_2, (3+1) via ungerade Rydberg states between 74500 and 90000 cm^{-1}].

A58. M. N. R. Ashfold, B. Tutcher, B. Yang, Z.-K. Jin, and S. L. Anderson, J. Chem. Phys. $\underline{87}$, 5105 (1987), [C_2H_2, (2+1) via gerade Rydberg states between 72400 and 77200 cm^{-1}].

A59. E. Sekreta, K. G. Owens, and J. P. Reilly, Chem. Phys. Lett. $\underline{132}$, 450 (1986), [Benzene, (1+1) via S_2 $^1B_{1u}$].

A60. K. S. Viswanathan, E. Sekreta, and J. P. Reilly, J. Phys. Chem. $\underline{90}$, 5658 (1986), [NO, (1+1) via A $^2\Sigma^+$, v=0].

A61. D. Normand, C. Cornaggia, and J. Morellec, J. Phys. B $\underline{19}$, 2881 (1986), [H_2, (4+2) via E,F, $^1\Sigma_g^+$, v_E=0, ATI].

A62. K. Sato, Y. Achiba, and K. Kimura, Chem. Phys. Lett. $\underline{126}$, 306 (1986), [$(NO)_2$, (1+1) near NO A $^2\Sigma^+$].

A63. B. G. Koenders, K. E. Drabe, and C. A. DeLange, Chem. Phys. Lett. $\underline{138}$, 1 (1987), [Br_2, (3+1) via Rydberg states].

A64. E. Y. Xu, T. Tsuboi, R. Kachru, and H. Helm, Phys. Rev. A (in press), [H_2, (3+1) via C $^1\Pi_u$, v=0-4].

A65. M. Lavolée, J. Kimman, J. W. J. Verschuur, and M. J. van der Wiel, in Photophysics and Photochemistry above 6 eV, edited by F. Lahmani (Elsevier, Amsterdam, 1985) p. 85, [NO (2+1) via C $^2\Pi$, B $^2\Pi$, and L $^2\Pi$].

A66. J. Kimman, J. W. J. Verschuur, M. Lavollée, H. B. van Linden van den Heuvell, and M. J. van der Wiel, J. Phys. B $\underline{19}$, 3909 (1986), [NO, (2+1') to high Rydberg states with v=3].

A67. L. A. Chewter, M. Sander, K. Müller-Dethlefs, and E. W. Schlag, J. Chem. Phys. $\underline{86}$, 4737 (1987), [Benzene, (1+1') ZEKE-PES].

A68. L. A. Chewter, K. Müller-Dethlefs, and E. W. Schlag, Chem. Phys. Lett. $\underline{135}$, 219 (1987), [Ar-benzene, (1+1') ZEKE-PES].

A69. J. W. Hepburn, D. J. Trevor, J. E. Pollard, D. A. Shirley, and Y. T. Lee, J. Chem. Phys. $\underline{76}$, 4287 (1982), [DF, (1+1) via B $^2\Delta$, v=2; CCl, 2 hν non resonant].

A70. R. J. S. Morrison, W. E. Conaway, T. Ebata, and R. N. Zare, J. Chem. Phys. $\underline{84}$, 5527 (1986), [NH_3, (2+1) via C' $^1A_1'$, v=0-9].

A71. J. R. Appling, M. G. White, T. M. Orlando, and S. L. Anderson, J. Chem. Phys. $\underline{85}$, 6803 (1986), [NO, (2+1') via A $^2\Sigma^+$, v=0].

A72. W. E. Conaway, T. Ebata, and R. N. Zare, J. Chem. Phys. $\underline{87}$, 3447 (1987), [NH_3, ND_3, (2+1) via C' $^1A_1'$, v=0-9].

A73. W. E. Conaway, T. Ebata, and R. N. Zare, J. Chem. Phys. $\underline{87}$, 3453 (1987), [NH_3, ND_3, (2+1) via C' $^1A_1'$, v=0-9].

A74. K. S. Viswanathan, E. Sekreta, E. R. Davidson, and J. P. Reilly, J. Phys. Chem. $\underline{90}$, 5078 (1986), [NO, (1+1) via A $^2\Sigma^+$, v=0; (2+1) via C $^2\Pi$ and D $^2\Sigma^+$, v=0].

A75. H. Sato, M. Kawasaki, K. Toya, K. Sato, and K. Kimura, J. Phys. Chem. $\underline{91}$, 751 (1987), [Trimethylamine, (1+1) via S_2].

A76. A. M. Woodward, S. D. Colson, W. A. Chupka, and M. G. White, J. Phys. Chem. $\underline{91}$, 274 (1987), [CH_3I, (2+1) via $^1\Sigma^+$, v=0, 1 and $^1\Pi$, v=0, 1].

A77. S. T. Pratt, P. M. Dehmer, and J. L. Dehmer, J. Chem. Phys. $\underline{87}$, 4423 (1987), [D_2, (3+1) via C $^1\Pi_u$, v=0-4].

A78. P. M. Dehmer and S. T. Pratt, J. Chem. Phys. (submitted), [Kr_2, (2+1) via Rydberg dissociating to $Kr+Kr^*5p[1/2]_0$].

A79. B. G. Koenders, D. M. Wieringa, K. E. Drabe, and C. A. DeLange, Chem. Phys. $\underline{118}$, 113 (1987).

PROGRESS IN RESONANCE ENHANCED MULTIPHOTON IONIZATION

SPECTROSCOPY OF TRANSIENT FREE RADICALS

by

Jeffrey W. Hudgens

Chemical Kinetics Division

Center for Chemical Physics

National Bureau of Standards

Gaithersburg, MD   20899

TABLE OF CONTENTS

I. INTRODUCTION

Resonance enhanced multiphoton ionization (REMPI) has become an important laser spectroscopic technique for studying transient free radicals. REMPI spectroscopy offers the advantage of excellent sensitivity -- even of radicals not suitably detected using laser induced fluorescence (LIF), Raman, or coherent antistokes Raman (CARS) spectroscopies. Although relatively new, its contributions to free radical spectroscopy are already significant. REMPI spectroscopy has provided a first view of the electronic spectra of several radicals and has doubled or tripled the spectroscopic data of others.

Spectroscopic data derived from REMPI studies have formed the basis for ultrasensitive detection schemes of radicals that play important roles during the etching of semiconductor surfaces, in gas phase atmospheric chemistry, and in combustion processes. Because REMPI apparatus is easily assembled from commercially available components and is adaptable to a wide range of experimental environments, research groups with diverse interests in various chemical processes have adopted REMPI spectroscopy to detect radicals produced during their experimental studies.

This review focuses upon the spectroscopic results which have accrued since the beginning of this active research field. The review is organized into three sections. The first presents background material to acquaint the nonexpert with the basic REMPI phenomenon. The second section describes typical apparatus and protocols used during REMPI spectroscopic studies of free radicals. The third section summaries the experimental results reported for specific free radicals.

II. AN OVERVIEW OF REMPI SPECTROSCOPY

A. A BRIEF HISTORY

The field of multiphoton ionization spectroscopy of free radicals has developed only recently. In 1975 the first REMPI spectroscopic results were reported by Johnson[1] at SUNY-Stony Brook who studied the stable radical, NO, and soon afterwards by Petty et al.[2] at the University of British Columbia who studied I_2. The first REMPI detection of a transient molecular free radical was reported in 1978 by Nieman and Colson[3] at Yale University. They observed a REMPI spectrum of NH (a $^1\Delta$) radicals produced by multiphoton photodissociation of NH_3. Two years later Colson's group also reported REMPI spectra of vibrationally excited NH_2 ($\tilde{X}$ $^2B_1$) and bands of NH (a $^1\Delta$) radical produced by UV/visible multiphoton photolysis of NH_3.[4]

Soon afterwards, the studies of methyl and trifluoromethyl radicals conducted by Hudgens' group at the Naval Research Laboratory demonstrated that REMPI spectroscopy could conveniently and very sensitively detect nonfluorescent free radicals. In 1981 DiGiuseppe et al. reported 3+1 REMPI spectra of methyl radicals produced by pyrolysis of various precursors in a tantalum foil oven. The REMPI spectrum appeared near 450 nm and originated from a three photon resonance with the 3d $^2E'' \leftarrow \leftarrow \leftarrow \tilde{X}$ $^2A_2''$ transitions.[5-6] In 1982 Duignan et al.[7] reported the 3+1 REMPI spectrum between 415-492 nm of CF_3 generated by pyrolysis in a tantalum oven and by infrared multiple photon dissociation (IRMPD).

In 1982 Prof. Welge's group at Universität Bielefeld reported detection of methyl radicals by using one laser to excite the 3s $^2A_1' \leftarrow \tilde{X}$ $^2A_2''$ transition and a second laser to ionize the excited 3s $^2A_1'$ methyl radicals generated by the first laser.[8] In studies published in 1982-83 Hudgens et al. reported observation of Rydberg states of methyl radical accessible only through multiphoton transitions.[9-10] This work reported the 3p $^2A_2'' \leftarrow \tilde{X}\ ^2A_2''$ two photon bands which are now commonly used to detect methyl radicals.

Since 1983 the REMPI spectroscopic data of free radicals have greatly expanded--to a large extent from research conducted at the National Bureau of Standards. Tables 1 and 2 show the current list of diatomic and polyatomic free radicals that have been studied with REMPI spectroscopy. The tables and this review include (currently) unpublished work from NBS and other groups. REMPI studies have revealed much new spectroscopic knowledge. For example, REMPI spectroscopy has led to the discovery of new electronic states in seventeen of the twenty-six radicals listed in Tables 1 and 2.

Table 1. Diatomic transient free radicals observed by REMPI spectroscopy.

Free Radical	Excitation Mechanism	Transition	Best Observation Wavelength (nm)	Reference
BrO	3+1	$F\ ^2\Sigma \leftarrow\leftarrow X\ ^2\Pi_{3/2}$	444.5	11
CCl	1+1	$A\ ^2\Delta \leftarrow X\ ^2\Pi$	271.4	12
CH	2+1	$D\ ^2\Pi_i \leftarrow\leftarrow X\ ^2\Pi_i$	310.8	13
ClO	3+1	$E\ ^2\Sigma \leftarrow\leftarrow X\ ^2\Pi_{3/2}$	437.8	11
CF	2+1	$D\ ^2\Pi \leftarrow\leftarrow X\ ^2\Pi_r$	346	14
NH $(a^1\Delta)$	3+1	$d\ ^1\Sigma^+ \leftarrow\leftarrow\leftarrow a\ ^1\Delta$	395.6	3
PH $(b^1\Sigma^+)$	2+1	$^3\Sigma^+ \leftarrow\leftarrow b\ ^1\Sigma^+$	388.9	15
PO	1+1	$B\ ^2\Sigma^+ \leftarrow X\ ^2\Pi_{3/2}$	320.4	16
SiF	1+1+1	$C''\ ^2\Sigma^+ \leftarrow A^2\Sigma^+ \leftarrow X^2\Pi_{1/2}$	437.5	17

Table 2. Polyatomic transient free radicals observed by REMPI spectroscopy.

Free Radical	Excitation Mechanism	Transition	Best Observation Wavelength (nm)	Reference
Triatomics				
CCO	3+1		443	18
NH_2	1+3	$\tilde{A}\ ^2A_1 \leftarrow \tilde{X}\ ^2B_1$	416.4	3
HCO	2+1	$3p\ ^2\Pi(A'') \leftrightarrow \tilde{X}\ ^2\Pi$	391.7	19
SiF_2	2+1	$\tilde{B}\ ^1B_2 \leftrightarrow \tilde{X}\ ^1A_1$	321.5	20
Polyatomics				
CH_3	2+1	$3p\ ^2A_2'' \leftrightarrow \tilde{X}\ ^2A_2''$	333.7	10
CF_3	3+1	$4s\ ^2A_1 \longleftrightarrow \tilde{X}\ ^2A_1$	455	7
CH_2F	2+1	$3p \leftrightarrow \tilde{X}\ ^2B_2$	378	21
CH_2OH	2+1	$3p \leftrightarrow \tilde{X}\ ^2A$	486.9	22
CH_3O	(a)		320.6	23
$CHCl_2$	2+1	$3d \leftrightarrow \tilde{X}\ ^2B_2$	370.1	24
Ethyl	2+1	$3p \leftrightarrow \tilde{X}\ ^2A_1$	410	25
Allyl	2+2	$3s\ ^2A_1 \leftrightarrow \tilde{X}\ ^2A_2$	498.8	26
Benzyl	2+2	$3\ ^2B_2 \leftrightarrow \tilde{X}\ ^2B_2$	502.5	27
Cyclohexanyl	(a)		500	28
2-methylallyl	2+2	$3s\ ^2A_1 \leftrightarrow \tilde{X}\ ^2A_2$	521.1	26
Cis-2-butene-1-yl	(a)		472.5	28
Trans-2-butene-1-yl	(a)		470	28

(a)Undetermined at the present time.

B. THE REMPI MECHANISM.

In nature the simultaneous absorption of more than one photon is a rare event. But when atoms and molecules are irradiated with the extreme intensity of a focused laser beam such as those generated by excimer and Nd:YAG pulsed dye lasers, simultaneous photon absorption rates greatly accelerate. A sufficiently intense laser can cause any molecule to simultaneously absorb enough photons to ionize. Under typical laser conditions ionization rates decrease rapidly with photon order, e.g., the simultaneous three photon absorption rate is much slower than the simultaneous two photon absorption rate; and the simultaneous four photon absorption rate is much slower than the three photon absorption rate.

The ionization rate of a chemical species is greatly enhanced when the path to ionization is divided into two or more successive absorption steps (resonances) of lower photon order. These absorption steps are provided by stable electronic states that can accumulate a population. For example, Figure 1 shows a simple diagram which depicts laser ionization of a chemical species by absorption of three laser photons. In Figure 1 the sum of two laser photons is resonant with an excited molecular state. This molecular state accumulates a population by simultaneous two photon absorption. Absorption of one more photon promotes the excited state molecule above its ionization potential and the molecule ionizes. The REMPI signal is detected by measuring the photoelectrons or laser generated cations.

The complete excitation process depicted in Figure 1 is called a 2+1 REMPI mechanism; two photon absorption populates a stable "resonant" electronic state and an additional one photon absorption step ionizes the molecule. Other laser ionization mechanisms, e.g., 3+1, 2+2, 1+3, in

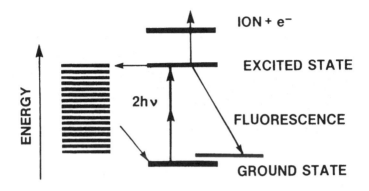

NONRADIATIVE
DECAY

Figure 1. Schematic of a 2+1 REMPI process. The competing processes, intramolecular relaxation and fluorescence, which deplete the excited state population and reduce the ion yield are also shown.

molecules have also been seen (Tables 1 and 2). In fact, within any n-photon ionization experiment which involves one stable resonant molecular state and one laser frequency the REMPI signal may arise from at least n-1 possible excitation schemes. This multiplicity of excitation mechanisms also presents a problem inherent to REMPI spectroscopy. Before spectroscopic constants can be derived, the experimentalist must determine the photon order of the resonant step which generates the REMPI spectrum.

To date, the REMPI spectra have only reflected the properties of the initial state and of the resonant intermediate excited state. The final absorption step which produces the ion does not appear to influence the spectrum strongly. This fact can be rationalized by recognizing that the ejection of the electron is usually very rapid (10^{-13}-10^{-15} sec). The Heisenberg uncertainty principle stipulates that the frequency width of the final state which lies above the ionization potential is very broad and relatively structureless.

Loss mechanisms which deplete the resonant excited state population before the ionizing photon is absorbed reduce the ion yield. Figure 1 shows some of these signal loss mechanisms which include 1) fluorescence from the resonant excited state to lower states, 2) collisionless and collisionally induced nonradiative relaxation of the resonant excited state into the dense background of vibrational levels associated with lower electronic states, and 3) predissociation into smaller neutral species. Since laser ionization occurs in less time than the laser pulse width (10^{-12}-10^{-8} sec), loss mechanisms cannot strongly attenuate the ion signal until their rates approach the inverse of the laser pulse duration. For example, evidence indicates that REMPI can overcome moderately rapid predissociation.[8,29]

In principle, because the excited state molecule exists just briefly before it absorbs an additional photon and ionizes, REMPI detection sensitivity should not rapidly decrease as pressure increases. Indeed, REMPI detection has demonstrated good sensitivity at sample pressures as high as one atmosphere.

C. MULTIPHOTON SELECTION RULES

As molecular symmetry increases, the number of excited states which are inaccessible to one photon spectroscopy also increases. Most of these "forbidden" states are accessible through two or three photon transitions. To help the experimentalist, several papers have listed selection rules,

rotational line strengths, and polarization factors for two, three, and four
photon absorption experiments on diatomic, symmetric, and asymmetric top
species.[30-41]

Because most free radicals studied to date are of low symmetry and few
have shown resolved rotational structure, this section concentrates on only
the most general multiphoton selection rules in the electric dipole
approximation. As needed, more specific rules are presented in later
sections.

In diatomic molecules governed by Hund's case (a) or (b) the one photon
absorption selection rule, $\Delta\Lambda=0$, ± 1, becomes $\Delta\Lambda=0$, ± 1, ± 2, ...$\pm n$ for n-
photon absorption. As the photon order of the resonant transition increases,
new states become accessible. For example, one photon absorptions from Σ
states permit the transitions, $\Sigma\leftarrow\Sigma$ and $\Pi\leftarrow\Sigma$. Simultaneous two photon
absorption permits the additional transition, $\Delta\leftarrow\Sigma$, and three photon
absorption adds the transition, $\Phi\leftarrow\Sigma$.

Multiphoton transitions give rise to more rotational branches. Whereas
one photon absorption bands may display P, Q, and R branches ($\Delta J=0$, ± 1), two
photon bands may show O, P, Q, R, and S branches ($\Delta J=0$, ± 1, ± 2) and three
photon bands may show N, O, P, Q, R, S, and T branches ($\Delta J=0$, ± 1, ± 2, ± 3).
The specifics of state symmetries, angular momentum coupling, and rotational
line strength factors may greatly simplify the spectrum by attenuating or
eliminating branches. Rotational branches often coincide. But in practice
the additional branches associated with multiphoton transitions often congest
REMPI bands sufficiently to preclude an extensive rotational analysis.

During standard derivations of the one photon vibrational selection
rules for polyatomic molecules, the Born-Oppenheimer approximation is invoked

to separate the electronic and nuclear wavefunctions. This separation permits independent solution of the electronic and vibrational interaction integrals which yield the electronic and vibrational selection rules. In similar fashion the application of the Born-Oppenheimer approximation permits division of the multiphoton interaction integrals into vibrational and electronic parts. With this approximation the vibrational interaction integral becomes identical to the one photon interaction integral. Thus, the same vibrational selection rules govern both one and multiphoton transitions.

The vibrational selection rules for polyatomic species have been described by Herzberg.[42] The strongest of these asserts that totally symmetric (e.g. a_i') vibrational modes follow the selection rule, $\Delta v=0, \pm 1, \pm 2, \ldots$ and that vibrational modes that are not <u>totally</u> symmetric are governed by the selection rule, $\Delta v=0, \pm 2, \pm 4 \ldots$

The data support use of the Born-Oppenheimer approximation. Vibrationally analyzed 2+1 and 3+1 REMPI spectra have been reported for three radicals (CH_3, CH_2F, $CHCl_2$) of C_{2v} symmetry or higher. The data span five different electronic band systems. In each band system symmetric modes and nonsymmetric modes were assigned and found to conform to these selection rules. Inevitably, exceptions to the approximate rules will be observed, but in these cases perturbation treatments of the vibration-electronic interaction will probably account for the spectra.'

D. CHARACTERISTICS OF RYDBERG STATES.[43]

Most radical electronic states observed by REMPI spectroscopy are Rydberg states. A Rydberg orbital is defined as a hydrogenic (s, p, d...) orbital that lies mostly outside of the valence shells of the atoms in the radical. The positions of Rydberg states conform to the equation:

$$\nu = IP - \frac{R}{(n - \delta)^2}$$

where IP is the adiabatic ionization potential of a cation state, R is the Rydberg constant (109,737 cm^{-1}), "n" is the principal quantum number of the excited electron, and δ is the quantum defect. The magnitude of the quantum defect depends upon the amount that the Rydberg orbital penetrates the molecular ion core. The values of δ vary but lie near specific values for each orbital type. For first row elements the values are $\delta(ns)\tilde{\,}1$, $\delta(np)\tilde{\,}0.6$, $\delta(nd)\tilde{\,}0.1$, and $\delta(nf)\tilde{\,}0.0$.[44-47] Generally the first member of a Rydberg series (usually n=3) is the most intense. The series members exhibit absorption intensities in proportion to $1/n^2$.

To a large extent the wavefunction of a Rydberg state molecule may be written:

$$\Psi = [\Phi_{cation}\Phi_{vib}\Phi_{rot}]\Phi_{Ryd}$$

where Φ_{cation} is the wavefunction of the cation core, Φ_{vib} is the vibrational wavefunction, Φ_{rot} is the rotational wavefunction, and Φ_{Ryd} is the hydrogenic Rydberg orbital wavefunction.

The brackets emphasize the fact that to a good approximation the wavefunction that describes the bonding and mechanical motion of the cation core is separable from the wavefunction that describes the Rydberg electron --particularly as "n" becomes large. Rydberg spectra reflect the separability of these wavefunctions. In most cases the vibrational frequencies observed in spectra of the Rydberg state closely resemble those of the cation.

III. EXPERIMENTAL ELEMENTS

A. APPARATUS

The experimental apparatus used for the multiphoton ionization studies of free radicals are conceptually simple. They are comprised of familiar components: a pulsed dye laser, an ion detector, a radical generation source, and the usual computer data acquisition or other signal processing equipment.

The Nd:YAG and excimer pumped dye lasers are identical to those used in laser excited fluorescence (LIF) experiments. In REMPI experiments most free radical spectra appear at wavelengths shorter than 580 nm. The photoexcitation conditions vary considerably. Typical laser parameters during experiments are: pulse durations between 8-20 nanoseconds and pulse energies between 0.1-20 mJ/pulse that are focused by a 38-250 mm focal length lens to irradiate a focal volume with an intensity between 0.4-22 gigawatts.

Ion detectors may be as simple as a pair of electrodes in a cell, but a quadrupole or time-of-flight mass spectrometer is particularly useful during exploratory spectroscopic studies of free radicals.

During REMPI spectroscopic studies numerous methods have been employed to produce free radicals. These methods include pyrolysis in ovens, infrared laser multiple photon dissociation (IRMPD), ultraviolet laser photolysis, bimolecular reactions in flow reactors, flames, and gas-surface heterogeneous reactions. Most new radical spectra have been discovered using bimolecular reactions in flow reactors.

Figure 2 shows a schematic of the experimental apparatus used at the National Bureau of Standards to study radicals. The flow reactor (adapted from a design by Anderson and Bauer[47]) depicted consists of three concentric glass tubes with diameters chosen so that the annuli formed between the tubes

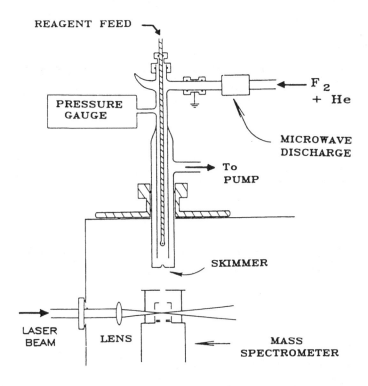

REAGENT FEED

PRESSURE GAUGE

F $_2$ + He

MICROWAVE DISCHARGE

To PUMP

SKIMMER

LASER BEAM

LENS

MASS SPECTROMETER

Figure 2. Schematic of the experimental zone of a REMPI mass spectrometer experiment which uses a flow reactor to generate free radicals.

are of equal area. A skimmer is mounted on the end of the largest tube. Equal

areas are required to obtain a constant flow velocity throughout the reactor.

Halogen atoms are prepared in a microwave discharge tube from halogen

molecules (F_2, Cl_2, Br_2) in helium. The discharge tube is connected to the

inner annulus of the flow reactor. A mixture of helium and halogen atoms

flows down the inner annulus to the end of the reactor where it reacts with a reagent injected by the smallest glass tube. A small portion of the reaction products pass through a skimmer and into the ion optics of the mass spectrometer where they are irradiated and ionized by a focused dye laser beam. The ions are mass selected, amplified, recorded, and signal averaged with the data acquisition system. The unsampled effluent exhausts though the outer annulus and through a vacuum pump.

The entire flow reactor is coated with halocarbon wax to inhibit radical and halogen atom recombination. Flow reactors are generally operated at pressures between 2-8 torr with molecular halogen/reagent/helium flow ratios of 2/4.4/5000 and a flow velocity between 300-3000 cm/sec. We estimate that the maximum radical density in the ion region of the apparatus is between 10^{10}-10^{11} radicals-cm^{-3}. When the radical density in the interaction zone is 5×10^{10} radicals-cm^{-3}, we estimate that the laser irradiates about 23,000 radicals.

B. IDENTIFICATION OF SPECTRAL CARRIERS

Most optical spectroscopies have difficulty confirming the identity of a spectral carrier when it is a free radical. In many cases the carrier is ascertained only after extensive rotational or vibrational analysis of its spectrum. Occasionally a spectrum has been erroneously attributed to a particular free radical when it actually originated from another transient species such as a vibrationally hot precursor, a stable product molecule, or a molecule in a metastable electronic state. As discussed below, with REMPI spectroscopy identification of a free radical spectral carrier is direct and fairly simple.

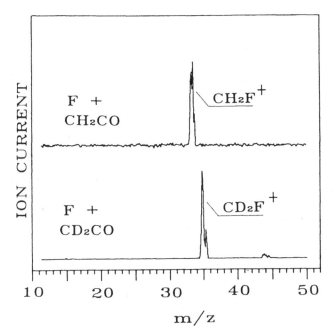

<u>Figure 3.</u> REMPI mass spectra of flow reactor effluent: a) F + Ketene at
359.05 nm. b) F + ketene-d_2 at 358.50 nm. (The small m/z 44 REMPI signal
originates from unreacted ketene-d_2.) From Ref. 21.

Reprinted with permission from J. Chem. Phys. in press, manuscript no. #A7.03.121.

1. The Unique Role of Mass Spectrometry.

The fact that REMPI signals are carried by ions which may be analyzed by
mass spectrometry provides unique opportunities for establishing the identity
of radicals and monitoring their production. This feature is particularly
valuable when the reaction of interest produces several products. For
example, the F + ketene reaction may proceed along three exothermic product
channels:

$$F(^2P) \quad + \quad H_2C=C=O \qquad \text{--->} \qquad CH_2F \quad + \quad CO$$
$$\text{--->} \qquad FHC=C=O \quad + \quad H$$
$$\text{--->} \qquad HC=C=O \quad + \quad HF$$

This reaction was recently studied in a flow reactor with REMPI spectroscopy. REMPI spectra were observed between laser wavelengths of 382-292 nm. Figure 3 shows typical mass spectra of the flow reactor effluent. Irradiation of the F + ketene reaction effluent generated m/z 33 ions (Figure 3a) and irradiation of F + ketene-d_2 effluent generated m/z 35 ions (Figure 3b). The shift of the REMPI spectrum by two atomic mass units showed that the transient species which carried the REMPI spectrum possessed two hydrogens. Thus, mass spectrometric evidence enabled an assignment of the m/z 33 and m/z 35 REMPI spectral carriers to the fluoromethyl radical.

More subtle structural assignments become possible by selective deuteration of the precursor. For example, the reaction of F-atoms with methanol produces both methoxy, CH_3O, and hydroxymethyl, CH_2OH, radicals. When Dulcey and Hudgens[48] conducted the F + CH_3OH reaction in a flow reactor, they observed a REMPI spectrum between 420-490 nm which was carried by m/z 31. To resolve between the two isomeric carriers, methoxy and hydroxymethyl radicals, they repeated the experiments using the reaction:

$$F \quad + \quad CH_3OD \quad \text{---->} \quad HF \quad + \quad CH_2OD \quad (m/z \ 32)$$
$$\text{---->} \quad DF \quad + \quad CH_3O \quad (m/z \ 31).$$

Figure 4 shows the results. Only the m/z 32 product was observed, which

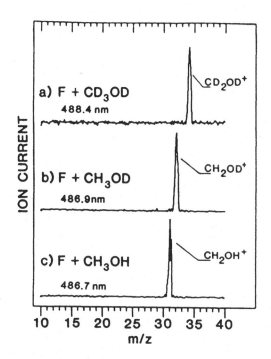

Figure 4. REMPI mass spectra of flow reactor effluent: a) F + CH₃OH, b) F + CH₃OD, and c) F + CD₃OD. From Ref. 48.

Reprinted with permission from J. Phys. Chem. 87, 2296 (1983) Copyright (1983) American Chemical Society

proved that the spectral carrier in this wavelength interval is the hydroxymethyl radical.

REMPI mass spectrometry is also useful for assaying the flow reactor conditions. For example, Figure 5a shows the flow reactor effluent when isobutene'was not reacted with halogen atoms. At 520 nm isobutene has a three

190

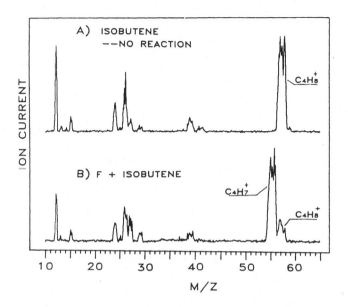

<u>Figure 5.</u> REMPI mass spectra of the flow reactor effluent produced by laser irradiation at 520 nm: A) isobutene--no reaction, B) F + isobutene. From Ref. 26.

Reprinted with permission from J. Phys. Chem. **89**, 1505 (1985) Copyright (1985) American Chemical Society

photon resonance with a 3p Rydberg state[49] which generates a REMPI mass spectrum carried by m/z 56 and several lower masses. (No m/z 55 signal appeared.) The relative isobutene concentration was measured by irradiating the sample with 520 nm laser light and measuring m/z 56 ion current.

When F atoms were added to the flow reactor and reacted with isobutene, the isobutene mass spectrum diminished and a large m/z 55 signal from the 2-

$$F \;+\; \underset{\underset{\displaystyle CH_3}{|}}{\overset{\overset{\displaystyle CH_3}{|}}{C}} = \underset{\underset{\displaystyle H}{|}}{\overset{\overset{\displaystyle H}{|}}{C} \quad \longrightarrow \quad HF \;+\; H_2C \overset{\overset{\displaystyle CH_3}{|}}{\overset{\displaystyle C}{\cdots}} CH_2 \qquad (m/z\ 55)$$

methylallyl radical appeared (Figure 5b). The increase in m/z 55 signal and
the depletion of m/z 56 signal was proportional to the amount of fluorine
introduced into the flow reactor. However, excessive fluorine or isobutene
suppressed the m/z 55 signal. By viewing the m/z 55 and m/z 56 signals as
functions of fluorine and isobutene concentration, the strongest 2-methylallyl
radical signal which also had the smallest m/z 56 signal from excess isobutene
was found.

These examples show some of the utility of mass spectrometry during REMPI
studies. Identification of spectral carriers by REMPI spectroscopy is direct.
Whenever an unknown REMPI spectrum is encountered, the associated mass
spectrum enables unambiguous identification of the molecular structure of the
laser ionized species. It should be noted that no other optical spectroscopy
possesses this advantage.

2. REMPI Mass Spectra of Free Radicals Show Little Fragmentation

The reported REMPI mass spectra of radicals rarely show significant
amounts of ion fragments. The mass spectra of radicals shown in Figures 3-6
are typical; they display only the molecular ions. Radicals from several
classes--ranging from the diatomic CF radical to large radicals like benzyl
and cyclohexanyl radicals (Figure 6)--show only the molecular ion. This

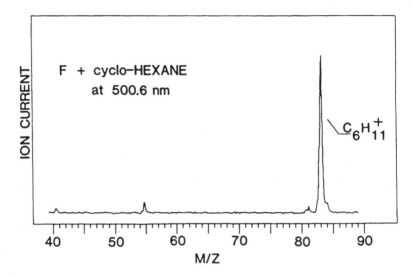

<u>Figure 6.</u> The REMPI mass spectrum of cyclohexanyl radical produced in a flow
reactor from the reaction: F + cyclohexane.

characteristic of laser ionization of radicals is intriguing and may be

typical for most radicals.

In contrast to free radicals, REMPI mass spectra of stable molecules

often display copious fragmentation, e.g. REMPI of isobutene shown in

Figure 5a. Fragmentation mechanisms observed during laser ionization were

recently reviewed by Gobeli et al.[50]

Each step of the REMPI process involving a stable molecule would seem to

guarantee abundant fragmentation. The REMPI process involving a stable

molecule usually removes a bonding electron. Diminished bonding can increase

the dissociative nature of the resonant state and cation. The cation formed

from stable molecules is a free radical. This radical cation is likely to possess a low lying valence state with dissociative character which lies in the same visible/UV wavelength interval as emitted by the laser. Such states provide a route for subsequent photolysis of the cation into ion fragments. In addition, the loss of a bonding electron upon ionization causes the cation to rearrange its geometry. The geometry change releases vibrational energy into the cation which also facilitates dissociation into daughter ions.

Most neutral free radicals possess an unpaired electron which resides in a nonbonding or antibonding orbital. In most cases laser ionization of radicals produces stable cations with closed electron shells. Cations which have closed electron shells do not usually have a low lying dissociative valence state within the laser wavelength range. Thus, closed shell molecular ions produced by REMPI of radicals are less likely to undergo subsequent photofragmentation into daughter ions. In addition, because the REMPI process usually removes a nonbonding or antibonding electron, the chemical bonding is enhanced or unchanged upon ionization. Geometry changes may occur but are less likely to release sufficient energy to bring about dissociation.

C. DETERMINATION OF THE PHOTON ORDER OF THE RESONANT STATE.

Since most multiphoton experiments have more than one possible excitation mechanism, a determination of the photon order of the resonant state is required of REMPI studies before spectroscopic constants may be derived. Experimentalists have used several approaches to prove the photon order of resonant transitions observed in free radicals. These approaches have included:

1) Measurement of the REMPI ion signal as a function of laser energy. Under ideal conditions [Ion Signal] $\propto I^n$ where "n" is the number of photons in the resonant step. Example: HCO radical.

2) Identification of ΔJ transitions in rotationally resolved REMPI spectra that indicate the photon order. Examples: CH_3, PO, PH(b $^1\Sigma^+$) radicals.

3) Analysis of excited state vibrational structure. Vibrational frequencies consistent with molecular structure are derived only when the proper photon order is assigned. Examples: CH_2OH, CH_2F, $CHCl_2$ radicals.

4) Identification of bands which reflect a spectroscopic property of the ground state such as spin-orbit splitting or vibrational hot bands. Examples: ClO, BrO, SiF radicals.

5) Analysis of Rydberg state 0_0^0 band shift vs. ionization potential for two or more similar compounds of the same symmetry. Examples: allyl, 2-methylallyl radicals.

6) Establishment of a Rydberg series. Reasonable solutions for Rydberg series are usually possible only when the proper photon order is assigned. Examples: CH_2F, CH_3 radicals.

7) Observation of the same electronic spectrum using two different photon orders. Examples: CF_3, benzyl radicals.

IV.   SUMMARY OF TRANSIENT RADICALS STUDIED BY REMPI SPECTROSCOPY.

In the next sections summaries of the REMPI spectroscopic studies of free

radicals are given. Each summary will describe the method in which the

radical carrier was proved, the REMPI mechanism, and principal spectral

information derived.

A.   DIATOMIC RADICALS

1. CF Radical

The CF radical has been extensively studied by many spectroscopic

techniques which have usually characterized the ground X $^2\Pi$ and lower lying

electronic states. The VUV spectroscopy of the higher lying Rydberg states of

CF was reported in a Ph.D. thesis by White[51] who analyzed several band systems

between 150-200 nm. This work is summarized in Huber and Herzberg's

compilation[52] and in a theoretical paper.[53]

Photoelectron spectroscopy studies of CF radical have reported an

adiabatic ionization potential of 9.11(0.01) eV.[54] The ionization process

eliminates an antibonding π-electron which increases the CF vibrational

frequency, ω_e, from 1308 cm^{-1} in the X $^2\Pi_r$ CF radical[54] to 1830(30) cm^{-1} in

the X $^1\Sigma^+$ cation.[55] Hepburn et al[56] have also reported the photoelectron

spectrum of nonresonantly multiphoton ionized CF radicals produced by

irradiation of CF_2Cl_2 and CCl_3F with 193 nm laser light.

Hudgens et al. observed m/z 31 REMPI spectra of CF radical between 340-

385 nm (Figure 7).[14] The CF radical was observed as a secondary reaction

product produced in the flow reactor during studies of the F + ketene and F +

CH_3F reactions. The CF radical spectrum displays strong REMPI bands centered

at 380.7, 368.2, 356.7, and 345.2 nm. Each band exhibits two bandheads spaced

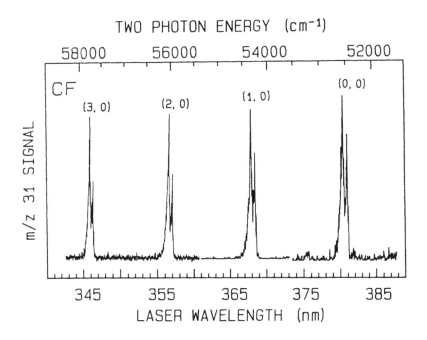

<u>Figure 7.</u> The REMPI spectrum of the CF radical between 343-385 nm.

Reprinted with permission from J. Chem. Phys. in press, manuscript no. #A7.03.121.

⁻0.5 nm apart. The resonant state was found to lie at the sum of two laser

photons (ν_{oo}⁻52,572 cm⁻¹) based upon the fact that the blue bandheads (at

380.43, 367.95, 356.5, 345.95 nm) fit a v'=0,1,2,3↔v"=0 vibrational series

with ω_e=1820 cm⁻¹ and $\omega_e x_e$=18.2 cm⁻¹. These values are very similar to the

ω_e=1840 cm⁻¹ and $\omega_e x_e$=20 cm⁻¹ observed in the CF (X $^1\Sigma^+$) cation.[54] This

similarity caused Hudgens et al. to assign the upper state of this band system

to a Rydberg state that possesses the same core configuration as the ground

state cation. In this view, the REMPI mechanism involves two photon

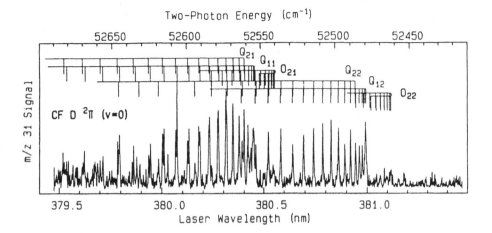

<u>**Figure 8.**</u>　The rotationally resolved 2+1 REMPI spectrum of the
CF (0,0) D $^2\Pi \leftrightarrow$ X $^2\Pi_r$ band.

Reprinted with permission from J. Phys. Chem., accepted and scheduled for publication
in Oct/Nov 87, manuscript # JP8708552-0-1-290

absorption to form the CF Rydberg state followed by absorption of one

additional photon to excite the Rydberg radical above the 9.11 eV[54] adiabatic

ionization potential, i.e. a 2+1 REMPI mechanism.

A rotational analysis of the CF v′=0,1,2↔v″=0 by Johnson and Hudgens[14]

has yielded more accurate spectroscopic constants (Table 3). Figure 8 shows

the rotationally resolved REMPI spectrum. The spectrum was expected to

display 20 rotational branches, but it is simplified by the small upper-state

spin splitting which makes certain pairs of branches degenerate. The most

prominent bands are the O- and Q-branches (ΔJ=0, -2). The results of this work

support an assignment of this band system as the D $^2\Pi \leftrightarrow$ X $^2\Pi_r$ transitions. The

present results concur with the spectroscopic constants and state assignment

derived by White[51] for a v′=1,2,3↔v″=0 VUV (one photon) spectrum.

Table 3. Spectroscopic constants for CF X $^2\Pi_r$ and D $^2\Pi$ states.

<u>Ground State Rotational Constants</u>

$\underline{v''=0}$

$B_v = 1.407332$ cm^{-1} Ref. 57.
$D_v = 6.62947 \times 10^{-6}$ cm^{-1} Ref. 57.
$A_v = 77.196916$ cm^{-1} Ref. 57, 58.

<u>D $^2\Pi$ State Rotational Constants</u>

$\underline{v'=0}$

$\nu_0 = 52,519.54 \pm 0.06$ cm^{-1}
$A_v = 6.8 \pm 0.3$ cm^{-1}
$B_v = 1.7197 \pm 0.0006$ cm^{-1}
$D_v = 7.1 \pm 1.5 \times 10^{-6}$ cm^{-1}

$\underline{v'=1}$

$\nu_0 = 54,298.94 \pm 0.05$ cm^{-1}
$A_v = 5.2 \pm 0.5$ cm^{-1}
$B_v = 1.6993 \pm 0.0008$ cm^{-1}
$D_v = 5.7 \pm 2.3 \times 10^{-6}$ cm^{-1}

$\underline{v'=2}$

$\nu_0 = 56,050.42 \pm 0.04$ cm^{-1}
$A_v = 6.8 \pm 0.3$ cm^{-1}
$B_v = 1.6814 \pm 0.0005$ cm^{-1}
$D_v = 4.5 \pm 1.0 \times 10^{-6}$ cm^{-1}

<u>Derived Constants for the D $^2\Pi$ State</u>

$T_e = 5,1619.37$ cm^{-1}
$\omega_e = 1,807.32$ cm^{-1}
$\omega_e x_e = 13.96$ cm^{-1}
$B_e = 1.7286 \pm 0.0005$ cm^{-1}
$\alpha_e = 0.0191 \pm 0.0005$ cm^{-1}

<u>Formulas</u>

$$T = T_e + G(v) + F(J)$$

$$G(v) = \omega_e (v+\tfrac{1}{2}) - \omega_e x_e (v+\tfrac{1}{2})^2$$

$$B_v = B_e - \alpha_e (v+\tfrac{1}{2})$$

Reprinted with permission from J. Phys. Chem., accepted and scheduled for publication in Oct/Nov 87, manuscript #JP8708552-0-1-290

2. CH Radical

The CH radical is commonly observed in combustion, interstellar, and laser induced chemistry. In the ground state the electronic configuration of the highest occupied orbitals is $2p\sigma^2 2p\pi^1$, X $^2\Pi_r$. Earlier UV absorption studies have thoroughly examined the three lowest excited valence states,[59-61] and the CH radical is routinely detected by LIF spectroscopy.[62] The spectroscopy of states lying above 50,000 cm^{-1} has been little studied.

In their report describing the first REMPI spectroscopy of the CH radical, Chen et al.[12] were able to solve a long-standing spectroscopic puzzle which had existed regarding in the band assignments. In 1969 Herzberg and Johns[63] reported the VUV spectrum of an nd Rydberg series and transitions terminating in three other states with origins at 58 981.0, 64 211.7, and 64 531.5 cm^{-1} which were designated D $^2\Pi_i$, E $^2\Pi_r$, and F $^2\Sigma^+$. Based upon the sign and magnitude of its spin-orbit coupling constant, the D $^2\Pi_i$ state was assigned the configuration of $2p\pi^3$. F $^2\Sigma^+$ was assigned to the $2p\sigma^2 3p\sigma$ Rydberg state based upon its position relative to the 3d Rydberg states. The puzzle which originated from this VUV study was: 1) What is the E-state configuration? 2) Why is no corresponding E-state band observed in the CD radical?

Chen et al.[13] observed the m/z 13 REMPI spectrum from CH radicals between 309-314 nm (Figure 9). The doubled output (1.0 mJ/pulse, 8 nsec) of a YAG pumped dye laser was focused with a 150 mm lens into a pulsed free jet expansion. Ketene and t-butylnitrite were photofragmented to produce CH radicals and photoionized by the same laser light pulse. The spectrum was

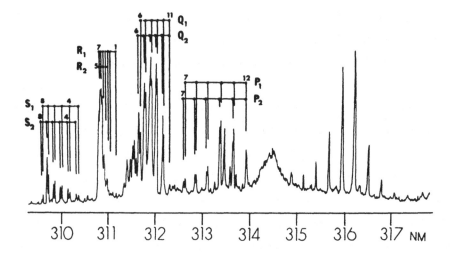

Reprinted with permission from J. Chem. Phys. 86, 516 (1986) Copyright (1986) American Institute of Physics

<u>Figure 9.</u> The 2+1 REMPI spectrum of CH radical (m/z 13) between 309-317 nm showing the rotationally resolved (2,0) D $^2\Pi_i \leftrightarrow$ X $^2\Pi_r$ band. From Ref. 65.

---

consistently assigned by assuming that the upper state resided at the sum of two photons. Rotational constants were calculated.

In Table 4 these new spectroscopic constants are compared with the older constants derived from the VUV spectrum. Of great significance is the spin-orbit coupling constant A. Because it is opposite in sign, but approximately the same in magnitude as that for X $^2\Pi_r$, Chen et al. assigned the E-state as the configuration of $2p\pi^3$.

These assignments based upon the REMPI spectrum solved the puzzle. The state previously assigned as the E $^2\Pi_i$ is in fact a vibrational member of the D $^2\Pi_i$ state. A reasonable vibrational frequency is obtained by assigning the band as the v'=2 level. The deuterium isotope shift of this vibrational level accounts for its absence in the CD radical spectrum.

Subsequent studies by Chen et al.[64,65] have presented voluminous data between 285-326 nm. In this work photoelectron spectra were recorded to help deconvolute contributions from overlapping and perturbing states. Table 5 shows new spectroscopic assignments derived from these studies.

Table 4. Spectroscopic constants of CH
D $^2\Pi_i$ (v'=2) radical.

Constant	REMPI Spectrum[a] (cm^{-1})	VUV Spectrum[b] (cm^{-1})
ν_o	64 209.4	64 211.7
B_v	12.7	12.6
D_v	15.8×10^{-4}	- - -
A	-26.2	- - -

[a]Ref. 13.

[b]Ref. 63.

Reprinted with permission from Chem. Phys. Lett. 121, 405 (1985) Copyright (1985) Elsevier Science Publishers BV

202

Table 5. Assignments and spectroscopic constants of CH states observed by 2+1 REMPI spectroscopy.[65] Values in parenthesis are tentative assignments.

State	Configuration	T_o, cm^{-1}	B_v, cm^{-1}	$10^4 D_v$, cm^{-1}
D $^2\Pi_i$	$2p\pi^3$	64 209.4	12.7	15.8
E' $^2\Sigma^+$				
v'=0	$2p\sigma^2 3p\sigma$	(63 000)		
v'=2		68 777.9	12.2	14.6
F' $^2\Pi$				
v'=0	$(2p\pi^2 3p\pi)$	(65 300)		
v'=1		(68 000)		
G' $^2\Sigma^+$				
v'=0	$(2p\sigma^2 4s\sigma)$	(69 900)		

Reprinted with permission from J. Chem. Phys. **86**, 516 (1986) Copyright (1986) American Institute of Physics

3. REMPI Spectra of ClO and BrO Radicals

ClO and BrO radicals are of great interest because they are central to the ClO$_x$, BrO$_x$ cycles of stratospheric ozone depletion. Although both radicals exhibit diffuse UV flame emission spectra,[66-69] they have no known LIF spectra. Early flash photolysis UV absorption studies had reported the lowest valence state[70] and six Rydberg states of ClO between 30,000-79,000 cm^{-1},[71] but of BrO only the lowest valence state between 22,000-28,000 cm^{-1} was known.[66] The REMPI study by Duignan et al.[11] reported the first spectrum of BrO between 54,000-72,000 cm^{-1} and reassigned the ClO F $^2\Sigma$ state origin.

Both radicals have ground state orbital configurations described by the terms:

$$\ldots (z\sigma)^2 (y\sigma^*)^2 (x\sigma)^2 (w\pi)^4 (v\pi^*)^3 \quad X \ ^2\Pi_i .$$

The lowest excited electronic states, A $^2\Pi_i$, are formed by a $\pi{\rightarrow}\pi^*$ transitions. The increased antibonding interaction in these A $^2\Pi_i$ states lowers the vibrational frequencies. The A $^2\Pi_i$ states display diffuse absorption bands. In contrast Rydberg states formed from $\pi^*{\rightarrow}R$ transitions. The promotion of the antibonding electron into a nonbonding Rydberg orbital increases the bonding and the observed vibrational frequency.

During the REMPI studies ClO and BrO were produced in a flow reactor by the reaction of halogen atoms with ozone:

$$X \ (X=Cl, \ Br) + O_3 \ \dashrightarrow \ O_2 + XO.$$

Chlorine and bromine atoms were produced in a microwave discharge. The laser linewidth used during these experiments was 0.8 cm^{-1}.

i. ClO

Figure 10 shows the REMPI spectrum of $^{35}Cl^{16}O$ between 417-474 nm. The bands observed in the spectrum are organized into six vibrational progressions. These progressions are further organized by pairing them into an intense progression and a weaker companion progression. The intensity ratio of the 0,0 bands in each progression pair is about 4.5:1 and the dye laser frequency separation is about 108 cm^{-1}. No rotational structure was resolved.

Using this evidence the REMPI spectrum of ClO was assigned to three photon resonances with high lying electronic states. The ground state of ClO is comprised of $^2\Pi_{1/2}$ and $^2\Pi_{3/2}$ levels separated by an A=-323 cm^{-1} spin-orbit coupling constant.[72,73] The Boltzmann population ratio of these spin-orbit states at 300K is about ~5:1 which accounts for the intensity ratios of the

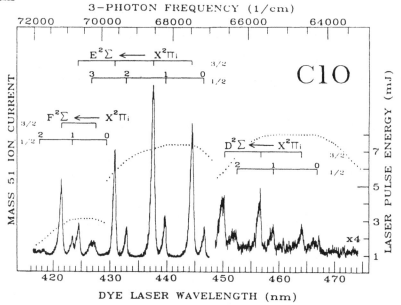

<u>Figure 10.</u> The three photon resonant REMPI spectrum of ^{35}ClO radical (m/z 51) between 416-474 nm. Dotted lines plot the average dye laser energy. From Ref. 11.

pairs of progressions. In REMPI spectra the dye laser frequency difference between each band pair is a submultiple of the ground-state spin orbit constant ($\sim$323 cm^{-1}). Since this interval is $\sim$108 cm^{-1} (i.e. 1/3 of the spin-orbit constant), the three photon nature of the transitions is proved. Thus, the REMPI structure of the REMPI spectrum arises from three photon resonances with Rydberg states which lie between 72,000-64,000 cm^{-1}.

Table 6 summarizes the REMPI bands reported by Duignan et al.[11] The agreement of this work is excellent except that the previously reported F(0,0) band[71] was not observed. This led Duignan et al.[11] to reassign the F-state band origin and vibrational numbering of the F $^2\Sigma$ state. This

Table 6. Transitions observed in ^{35}ClO by 3+n REMPI spectroscopy.[11]

TRANSITION (ν',ν'')Ω''	Dye Laser Wavelength (nm)	Three Photon Energy (cm^{-1})	Vibrational Interval (cm^{-1})
D $^2\Sigma \longleftrightarrow X\ ^2\Pi_i$			
(0,0) 3/2	464.28	64 598	0
(1,0) 1/2	459.13	65 323	- - -
3/2	456.87	65 646	1 048
(2,0) 1/2	452.53	66 275	952
3/2	450.32	66 601	955
E $^2\Sigma \longleftrightarrow X\ ^2\Pi_i$			
(0,0) 1/2	446.92	67 107	0
3/2	444.73	67 438	0
(1,0) 1/2	439.84	68 188	1 081
3/2	437.78	68 508	1 070
(2,0) 1/2	433.15	69 241	1 053
3/2	431.14	69 563	1 055
(3,0) 1/2	(426.80)	(70 271)	1 030
3/2	424.81	70 600	1 037
(4,0) 3/2	418.63	71 642	1 042
F $^2\Sigma \longleftrightarrow X\ ^2\Pi_i$			
(0,0) 1/2	(429.42)	69 842	0
3/2	427.49	70 157	0
(1,0) 1/2	423.51	70 817	975
3/2	421.65	71 129	972
(2,0) 1/2	(417.78)	(71 788)	971

reassignment was justified on the basis that the REMPI spectrum viewed vibrationally cool radicals whereas the VUV flash photolysis study viewed the collective spectrum of very vibrationally hot radicals, precursors, and perhaps secondary reaction products. Because the m/z 51 REMPI spectrum recorded principally signal from the ClO^+ molecular ion, contributions from precursors and secondary products were strongly discriminated against.

The vibrational frequencies, ω_e, observed among the excited states range between 981-1085 cm^{-1}. These frequencies resemble the 1040(30) cm^{-1} vibrational frequency observed in ClO^+ $(X\ ^3\Sigma^-)$.[74,75] This similarity supports the assignment of the upper states to Rydberg states which possess a $X\ ^3\Sigma^-$ cation core.

Mass spectra of the ClO radical showed that the REMPI signal was comprised of 90% ClO^+ and 10% Cl^+. The Cl^+ signal mimicked the ClO^+ signal, indicating that the fragmentation occurred after the three photon states were prepared.

In the REMPI spectrum the D-state progressions are much weaker than the E-state progressions. Duignan et al. accounted for the weaker D-state intensity by noting that the 10.87 eV[74,75] adiabatic ionization potential of ClO is equal to four photons of 456 nm light. The radicals ionized by laser wavelengths longer than 456 nm must absorb five photons. As a result D-state progressions are observed through a 3+2 REMPI mechanism and the E-state progressions are observed through a 3+1 REMPI mechanism. Thus, the lower intensity of the D-state compared to the E-state probably reflects the lower two photon cross-section at the ionization step.

Table 7 summarizes the spectroscopic constants of ClO obtained from REMPI and absorption spectroscopies.

Table 7. Spectroscopic constants for states of ^{35}ClO and ^{79}BrO.

State	T_e, cm^{-1}	ω_e cm^{-1}	$\omega_e x_e$, cm^{-1}	$-A$, cm^{-1}	References
ClO					
H	74 132[a]	$\Delta G_{1/2} = 1023$			71
G $^2(\Lambda=1,2)_i$	73 600[a]	1 066	5	39[a]	71
F $^2\Sigma$	70 093	981	1.4	0	11
E $^2\Sigma$	67 323	1 085	3.6	0	11
D $^2\Sigma$	64 500	1 052	3	0	11, 71
C $^2\Sigma^-$	58 490	1 077.74	6.365	0	73
A $^2\Pi_i$	31 750	519.5	7.2	521[a]	79, 80
X $^2\Pi_i$	0	853.72	5.58	321.8	72, 73
BrO					
G $^2(\Lambda=1,2)_i$	70 441	$\Delta G_{1/2} = 848$		139[a]	11
F $^2\Sigma$	67 420	(822)[d]		0	11
E $^2\Sigma$	64 916	897	2.6	0	11
A $^2\Pi_{3/2}$	26 098	511.3	4.83	----	81
X $^2\Pi_i$	0	752.54	4.932	968.0	76, 77

[a] Spectroscopic T_e and $-A$ constants are revised to conform to the X $^2\Pi_i$ constants of Refs. 72 and 73.

[b] Spectroscopic T_e constants are derived based upon the X $^2\Pi_{3/2}$ constants from Refs. 76 and 77.

[c] For $^2\Pi_i$ assignment.

[d] Average interval between $v'=2-0$.

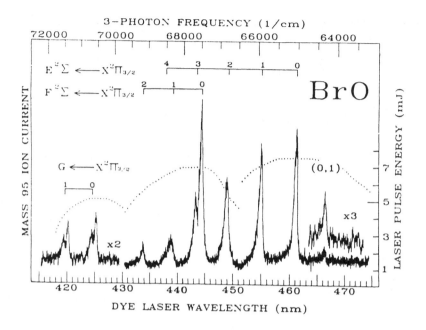

Figure 11. The 3+1 REMPI spectrum of ^{79}BrO radical (m/z 95) between 415-475 nm. Dotted lines plot the average dye laser energy. From Ref. 11.

Reprinted with permission from J. Chem. Phys. **82**, 4426 (1985) Copyright (1985) American Institute of Physics

ii. BrO

The laser induced ion current from BrO radical was comprised of 90% molecular ion BrO$^+$ and 10% Br$^+$; behavior which parallels that of ClO.

The REMPI spectrum of ^{79}Br^{16}O in the region 417-474 nm is shown in Figure 11. No rotational structure was resolved. Table 8 lists positions of the bandheads. These bands are organized into three progressions. No doublet structure originating from the ground state $^2\Pi_{1/2}$--$^2\Pi_{3/2}$ spin-orbit splitting was observed, nor was it expected since the ground state spin-orbit splitting in BrO is A=-968 cm^{-1},[76] much greater than in ClO. The upper spin-orbit state, $^2\Pi_{1/2}$, was sparsely (<1 %) populated at 300 K which

Table 8. Transitions observed in ^{79}BrO by 3+1 REMPI spectroscopy.

TRANSITION $(\nu',\nu'')\Omega''$	Dye Laser Wavelength (nm)	Three Photon Energy (cm^{-1})	Vibrational Interval (cm^{-1})
$E\ ^2\Sigma \longleftrightarrow X\ ^2\Pi_{3/2}$			
(0,1)	466.51	64 289	-714
(0,0)	461.39	65 003	0
(1,0)	455.17	65 891	888
(2,0)	449.22	66 764	873
(3,0)	443.49	67 626	862
(4,0)	437.89	68 491	865
$F\ ^2\Sigma \longleftrightarrow X\ ^2\Pi_{3/2}$			
(0,0)	444.52	67 470	0
(1,0)	439.25	68 279	809
(2,0)	433.94	69 115	836
$G\ ^2(\Omega = |\Lambda| + 1/2) \longleftrightarrow X\ ^2\Pi_{3/2}$			
(0,0)	425.39	70 504	0
(1,0)	420.33	71 352	848
$G\ ^2(\Omega = |\Lambda| - 1/2) \longleftrightarrow X\ ^2\Pi_{3/2}$			
(0,0)	424.54	70 645	0
(1,0)	419.53	71 489	844

thwarted an assignment of the photon order of the resonant step on the basis of the observed doublet energy spacing.

Duignan et al. determined the photon order of the resonant states by noting a hot band at 466.51 nm. The laser energy difference between this band and the origin, E(0,0), was 1/3 of the known ground state vibrational frequency[77] which showed that the upper electronic state resides at the sum of three laser photons.

The vibrational intervals among the three progressions vary between 809-888 cm^{-1} which are 80-160 cm^{-1} greater than observed in the ground state. The similarity of the upper resonant states to the 830(30) cm^{-1} vibrational frequency observed in BrO^+ $(X\ ^3\Sigma^-)$[78] led Duignan et al. to assign the upper E and F states to Rydberg states which possess a $X\ ^3\Sigma^-$ cation core.

The vibrational progression between 425-415 nm, labeled G, is composed of a pair of doublets. The first member and apparent origin of the G-series lies at 425.39 nm ($3h\nu$=70 504 cm^{-1}). The doublet splitting within both bands is 139 cm^{-1}.

Because of the doublet structure Duignan et al. assigned these bands to a state possessing nonzero orbital angular momentum. They proposed that the G-state in BrO has an electron configuration similar to the G-state reported in ClO by Basco and Morse.[71] The G-state term values in both radicals are similar. However, neither the REMPI data for BrO nor the VUV data for ClO lead to a firm state assignment. The VUV and REMPI data only limit possible assignments for the G-states in ClO and BrO to either $^2\Pi_i$ or $^2\Delta_i$.

The observation of doublet structure in BrO shows a fundamental difference in the selection rules which govern multiphoton spectroscopy. A one photon spectrum of the same upper state would have displayed only singlet bands. This dissimilarity in appearance between one and multiphoton spectra is understood by examining the governing selection rules while keeping in mind that only the lower spin-orbit level of BrO, X $^2\Pi_{3/2}$, is populated.

For Hund's case (a) molecules the one photon selection rules are $\Delta\Lambda=0$, ± 1 and $\Delta\Sigma=0$; n-photon selection rules follow $\Delta\Lambda=0$, $\pm 1,\ldots\pm n$ and $\Delta\Sigma=0$. Σ is the projection of spin upon the internuclear axis ($\Omega=|\Lambda + \Sigma|$). Thus, in a one photon VUV absorption spectrum to BrO upper states of $^2\Pi$ or $^2\Delta$ symmetry only one band ($^2\Pi_{3/2}\leftarrow X^2\Pi_{3/2}$ or $^2\Delta_{5/2}\leftarrow X^2\Pi_{3/2}$) will be observed. In contrast three photon resonant REMPI spectrum will display the additional bands, $^2\Pi_{1/2}\leftrightarrow X\ ^2\Pi_{3/2}$ and $^2\Delta_{3/2}\leftrightarrow X\ ^2\Pi_{3/2}$. Thus, REMPI spectra of $^2\Pi$ or $^2\Delta$ states should display doublet bands. (Three photon transitions to $^2\Gamma_{9/2}$ and $^2\Phi_{7/2}$ states are also allowed but will only produce singlet bands.)

A two photon experiment which measures these G-states could lead to a conclusive assignment. In this case the additional $^2\Pi_{3/2}\leftrightarrow X\ ^2\Pi_{3/2}$ band is allowed, but the $^2\Delta_{3/2}\leftrightarrow X\ ^2\Pi_{3/2}$ band is forbidden. Thus, a $^2\Pi\leftrightarrow X^2\Pi_{3/2}$ band system will display doublet bands and a $^2\Delta\leftrightarrow X^2\Pi_{3/2}$ band system will display only singlet bands. The two photon spectrum between 276-287 nm needed to resolve this assignment has not been reported.

4. CCl Radical

Sharpe and Johnson[12] have observed a REMPI spectrum of supersonically cooled CCl radicals between 271-271.5 nm. The CCl radicals were produced in a supersonic expansion of helium and CCl_4 by subjecting the mixture to an electric spark at the exit aperture of a pulsed valve. In their apparatus a time-of-flight mass spectrometer detected the laser generated ions. The laser bandwidth used to produce the spectrum was 0.2 cm^{-1} FWHM.

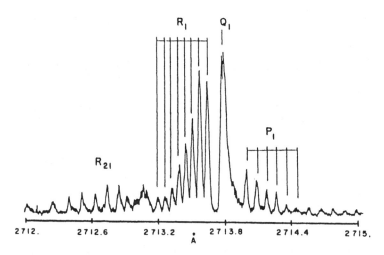

<u>Figure 12.</u> The 1+1 REMPI spectrum between 271-271.5 nm of rotationally cold (30K) $C^{35}Cl$ radical produced by a pulsed electric discharge in pulsed supersonic expansion of Ar/CCl_4. From Ref. 11.

The rotationally resolved REMPI spectrum of $C^{35}Cl$ radicals is shown in Figure 12. Analysis of the spectrum showed that it arises from A $^2\Delta(v'=1)\leftarrow X$ $^2\Pi_{1/2}(v''=0)$ one photon transitions and has a rotational temperature of 30K. Since the ionization potential of CCl is 8.9(2) eV,[56] the radical may ionize through an autoionizing level after absorbing one more photon, i.e. a 1+1 REMPI mechanism.

5. NH (a^1 Δ) Radical

The NH (a $^1\Delta$) radical was the first transient radical detected by REMPI spectroscopy. Nieman and Colson[3] observed a REMPI spectrum of the NH ($a^1\Delta$) radical between 394.5-397 nm. They attributed the NH radical production

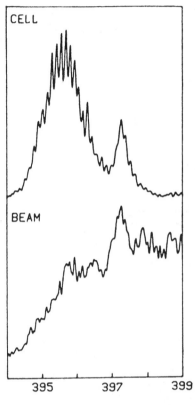

CELL

BEAM

395   397   399

LASER WAVELENGTH (NM)

Figure 13. The REMPI spectra observed between 394-399 nm from the irradiation of NH_3 in a static cell (upper panel) and within a molecular beam (lower panel). Ref. 4.

Reprinted with permission from J. Chem. Phys. 73, 4296 (1980) Copyright (1980) American Institute of Physics

mechanism to three photon photodissociation of NH_3. These experiments were conducted in a laser ionization cell which contained a static fill of 5 torr NH_3 gas. A nitrogen pumped dye laser was used to generate the REMPI signals.

In a later paper Glownia et al.[4] repeated these experiments between 394-397 nm in a supersonically cooled molecular beam. No NH radical spectrum was

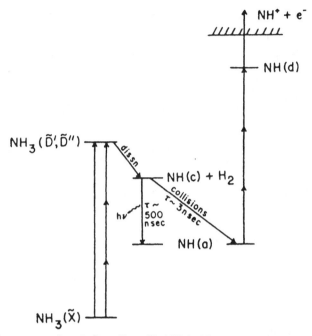

Figure 14. Schematic representation of the photophysical processes caused by irradiation of NH_3 by an intense laser beam tuned between 394-397 nm. From Ref. 4.

---

observed in the supersonically expanded gas, but an NH spectrum was observed in the static cell (Figure 13). Glownia et al. accounted for the production of NH spectra via the series of processes shown in Figure 14. Between 395-397 nm three photon absorption prepares NH_3 ($\tilde{D}'$ $^1A_2''$ and $\tilde{D}''$ $^1A_2''$) which rapidly dissociates to form NH (c $^1\Pi$) + H_2 ($^1\Sigma_g$). In a static cell at 5 torr a fraction of the nascent NH (c $^1\Pi$) product is collisionally quenched within the duration of the 5 nsec laser pulse to form NH (a $^1\Delta$). The NH (a $^1\Delta$) is ionized via a 3+1 REMPI mechanism through the three photon resonant d $^1\Sigma^+$ state.

The model proposed by Glownia et al.[4] also accounts for the absence of a NH (a $^1\Delta$) REMPI spectra under molecular beam conditions. In a supersonic beam collisional relaxation of NH (c $^1\Pi$) into NH (a $^1\Delta$) does not occur. Instead, NH (a $^1\Delta$) forms more slowly through radiative decay ($t_o \sim 500$ nsec) of NH (c $^1\Pi$). Thus, no REMPI signal from NH (a $^1\Delta$) was seen because an insufficient concentration of NH (a $^1\Delta$) was formed during the duration of the 5 nsec laser pulse.

6. PH (b $^1\Sigma^+$) Radical

Ashfold et al.[15] have examined the spectra of PH (b $^1\Sigma^+$) and PD (b $^1\Sigma^+$) produced by photolysis of PH_3 and PD_3. These experiments were conducted in an MPI cell comprised of evacuable chamber equipped with entrance and exit windows for the laser beam and two parallel platinum plates biased at ~ 100 VDC. About 20 torr of PH_3 or PD_3 was introduced into the cell. A pulsed laser was focused into the cell and the REMPI ions were collected by the parallel plates, amplified, and processed by a boxcar integrator prior to display on a strip chart recorder.

Ashfold et al. proposed that the PH (b $^1\Sigma^+$) radicals were generated by simultaneous three photon absorption into a dissociative state of PH_3. In support of this mechanism they observed that at the lowest laser energies (0.5-1.2 mJ/pulse) the REMPI signal intensity increased in proportion with the cube of the laser power.

Ashfold et al.[15] reported REMPI spectra observed between 383-394 nm. A strong Q-branch structure in PH_3 was centered at ~ 388.9 nm. This Q-branch decreased in intensity as the laser polarization changed from linear to circular. This dependence of the Q-branch intensity upon laser polarization showed the multiphoton nature of the band. Rotational structure from O and S branches was analyzed for both PH and PD radicals (Table 9). The structure of this band originates from a two photon resonance with the $^3\Sigma^+ \leftarrow\leftarrow b\ ^1\Sigma^+$ transition. The $^3\Sigma^+$ state is of the configuration $KL(3s\sigma)^2(3p\sigma)^2(3p\pi)^1(4p\pi)^1$. Since the ionization potential of PH radical is 10.19,[52,82] ionization proceeds through a 2+1 REMPI mechanism.

Ashfold et al. pointed out that the observed singlet-triplet transition was possible because of mixing of triplet character (probably from A $^3\Pi$) with the b $^1\Sigma^+$ state. Evidence for a second mixing mechanism which relaxes the selection rules, $\Delta S=0$ and $0^- \nleftrightarrow 0^+$ was also presented.

Table 9. Spectroscopic constants for $b^1\Sigma^+$ and $^3\Sigma^+$ states of PH and PD reported in Ref. 15. X $^3\Sigma^-$ rotational constants are from Ref. 83.

State	T_0 (cm^{-1})	B_0 (cm^{-1})	$10^4 D_0$ (cm^{-1})
PH			
$^3\Sigma^+$	$a_{PH} + 51431.6 \pm 1.0$	$8.40_4 \pm 0.01$	$4.3_5 \pm 0.2$
b $^1\Sigma^+$	a_{PH}[a]	$8.45_9 \pm 0.01$	4.1 ± 0.2
X $^3\Sigma^-$	0	8.4114	4.34
PD			
$^3\Sigma^+$	$a_{PD} + 51433.5 \pm 1.0$	$4.36_4 \pm 0.01$	$1.2_8 \pm 0.1$
b $^1\Sigma^+$	a_{PD}[a]	$4.40_5 \pm 0.01$	1.2 ± 0.1
X $^3\Sigma^-$	0	4.3617	1.16

[a] $a_{PH} = 14345.2 \pm 0.2$ cm^{-1} from Ref. 84.

7. PO Radical

The PO radical is of interest to the combustion scientist because of its role in inhibition and promotion processes which occur in premixed and diffusion flame containing phosphorus additives.[85] It has a well known optical spectrum[52] and may be detected using laser excited fluorescence.[86]

Smyth and Mallard[16] observed the REMPI spectrum of PO X $^2\Pi$ radical in a fuel rich, premixed acetylene/air flame operated in an open slot burner. A schematic of this apparatus is shown in Figure 15. The ionization signals were produced by irradiation of the flame with a focused (fl=150 mm), 100 μJ/pulse doubled dye laser beam. The laser generated electrons were

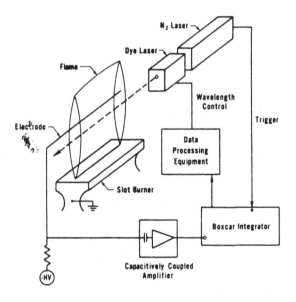

Figure 15. Schematic diagram of the apparatus for ionization studies in flames. From Ref. 16.

collected with a tungsten electrode inserted into the flame just above the laser focus. The laser generated electrons were amplified and averaged by a boxcar averager.

In a flame PO radical arises from the oxidation of the phosphine impurity, PH_3, which is present in commercial acetylene at a mole fraction of $\sim 10^{-5}$. The same REMPI spectrum was observed when triphenylphosphine was added to ethylene/air flames.

A REMPI spectrum was observed between 302-334 nm which displayed a series of bandheads (Figure 16). These bandheads corresponded to bands detected in the one-photon emission spectrum for PO B $^2\Sigma^+ \leftarrow$ X $^2\Pi_{1/2}$ and B $^2\Sigma^+ \leftarrow$ X $^2\Pi_{3/2}$ transitions. Assignment of the spectral features was based upon previously reported PO spectroscopic constants.[87-89] In addition to the B $^2\Sigma^+ \leftarrow$ X $^2\Pi_{1/2,3/2}$ transitions, four bandheads were detected in the 262-264 nm region which could be assigned to A $^2\Sigma^+ \leftarrow$ X $^2\Pi_{1/2,3/2}$ transitions: (0-2) 38,151; 37,947; and 37,926 cm^{-1} and (1-3) 38,110 cm^{-1}.

After the first laser photon prepared the PO A $^2\Sigma^+$ or B $^2\Sigma^+$ states, the ionization of these excited PO radicals was believed to occur through two different sequences: 1) When the sum of two photons exceeded the ionization potential, direct one photon absorption ionized the excited state PO radicals. 2) When the sum of two laser photons was less than the ionization potential, the second photon absorption prepared electronic states of PO that resided near the ionization onset energy. Subsequent collisions ionized the highly excited PO radicals.

Based upon results obtained with a 100 μJ/pulse laser, Smyth and Mallard reported that at 2000K REMPI spectroscopy could detect about 3×10^{10} radicals-cm^{-3} in an open flame operating at 1 atm.

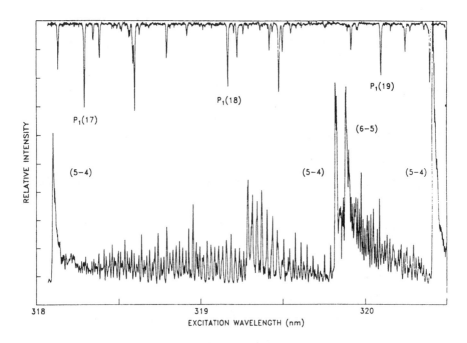

<u>Figure 16.</u> Experimental 1+1 REMPI spectrum of PO radical (lower trace) and the LIF spectrum of OH radical obtained simultaneously for wavelength calibration (upper trace). The labeled (v',v") bandheads for PO arise from $B^2\Sigma^+ \leftarrow X\ ^2\Pi$ transitions. The feature at 319.5 nm has not been identified. From Ref. 16. Reprinted with permission from J. Chem. Phys. **77**, 1779 (1982) Copyright (1982) American Institute of Physics

Smyth and Mallard computed relative REMPI cross-sections by dividing the signals by the lower state Boltzmann population and by the Franck-Condon factor. The results of these calculations lead Smyth and Mallard to believe that some REMPI bands possessed anomalously high ionization cross-sections. To explain the anomalous intensity behavior, they proposed that the ionization continuum of PO also possessed structure originating from autoionizing levels. They warned that their conclusions were tentative until more accurate RKR potential curves for the Franck-Condon calculation became available--particularly for the smaller Franck-Condon factors.

Wong et al.[86] subsequently recalculated the Franck-Condon factors; but their values differ little from those of Smyth and Mallard. They accounted for the anomalously large (5,4) and (6,5) bandhead intensities by using the R-centroid approximation to calculate the electronic transition moment. In summary, it is not clear from these studies that the role of autoionizing states selected by the absorption of the second photon has been resolved.

8. SiF Radical

The SiF radical is of interest to the semiconductor industry since it is one of the products observed during chemical etching reactions at silicon surfaces. SiF radical may be detected by LIF of the A $^2\Sigma^+$ state around 440 nm.[90] Previous studies of its UV spectrum between 40,000-48,000 cm^{-1} were reported by Barrow et al.[91] and by Houbrechts et al.[92]

Dulcey and Hudgens[17] have reported the REMPI spectrum of SiF between 430-480 nm (Figure 17). In the REMPI study SiF was produced in a flow reactor by the reaction of a large excess of fluorine atoms with SiH$_4$ at ~1.9 torr. This reaction was ill-characterized, but by assuming 100% conversion of SiH$_4$ to SiF radicals, they estimated that a maximum concentration of 5×10^{10} SiF radicals-cm^{-3} resided in the laser photoionization zone.

Only SiF$^+$ ions (m/z 47) carried the REMPI spectrum. The spectrum shows two different vibrational progressions of doublets which originate from two different upper electronic states. The doublet structure arises from the spin-orbit coupling in the ground state which separates the X $^2\Pi_{3/2}$ and X $^2\Pi_{1/2}$ states by 162 cm^{-1}.[93-94] The 2:1 doublet intensity ratios reflect

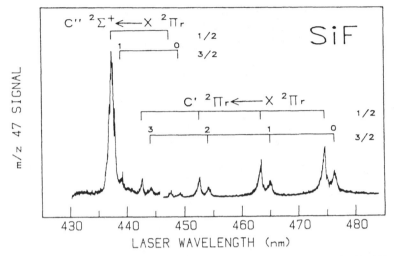

<u>Figure 17.</u> The REMPI spectrum of SiF radical (m/z 47) between 430-484 nm.
From Ref.17.

their 2:1 Boltzmann population ratio. Each doublet is separated by about 79
cm^{-1} of dye laser (one photon) energy. Since this interval is nearly one
half of the ground state spin orbit coupling constant, it provided direct
evidence that these doublet bands originate from two photon transitions.
The ionization potential of SiF is 7.28 eV.[95] Three photons are required to
ionize SiF, i.e. a 2+1 REMPI mechanism.

The assignments of the REMPI bands observed by Dulcey and Hudgens are
listed in Table 10. The agreement of these two photon bands with the one
photon spectrum reported by Houbrechts et al.[85] supported the assignment of
the REMPI spectrum to C' $^2\Pi_r \leftrightarrow$ X $^2\Pi_r$ and C" $^2\Sigma^+ \leftrightarrow$ X $^2\Pi_r$ transitions.

The (1,0) C" $^2\Sigma^+ \leftrightarrow$ X $^2\Pi_r$ doublet pair exhibit an anomalously large 7:1
intensity ratio. This anomalous REMPI band intensity was accounted for by
noting that the one photon SiF (0,0) A $^2\Sigma^+ \leftarrow$ X $^2\Pi_{1/2}$ band lies at 436.8 nm[90]

Table 10. The 2+1 REMPI bands observed in SiF radical.[17]

Transition $(v',v'')\Omega''$	Laser Wavelength (nm, air)	Two Photon Frequency (cm^{-1})	Vibrational Interval (cm^{-1})
C' $^2\Pi_r \leftrightarrow$ X $^2\Pi_r$			
(0,0) 3/2	476.3	41 979	0
1/2	474.6	42 138	0
(1,0) 3/2	465.0	42 999	1 020
1/2	463.3	43 157	1 019
(2,0) 3/2	454.2	44 021	1 022
1/2	452.6	44 177	1 020
(3,0) 3/2	444.1	45 022	1 001
1/2	442.6	45 175	998
C" $^2\Sigma^+ \leftrightarrow$ X $^2\Pi_r$			
(0,0) 3/2	449.3	44 502	0
1/2	447.7	44 661	0
(1,0) 3/2	439.1	45 535	1 033
1/2	437.5[a]	45 701	1 040

[a] Enhanced at the one photon level by the A $^2\Sigma^+ \leftarrow$ X $^2\Pi_{1/2}$ transition at 436.8 nm which was measured in the LIF study of Ref. 90.

and overlaps with the band center of the anomalously large two photon (1,0) C" $^2\Sigma^+ \leftrightarrow$ X $^2\Pi_{1/2}$ band at 437.5 nm. The REMPI cross-section became larger because the REMPI mechanism changed from a 2+1 process into a sequence of one photon transitions between long-lived states, i.e. a 1+1+1 REMPI mechanism.

For sequential 1+1+1 REMPI mechanisms to be important, connecting one
photon steps along the path to ionization must lie within the laser
bandwidth. For example, the (1,0) C" $^2\Sigma^+\leftrightarrow$X $^2\Pi_{3/2}$ transition at 439.1 nm
showed no enhancement of the REMPI cross-section because the connecting one
photon (0,0) A $^2\Sigma^+\leftarrow$X $^2\Pi_{3/2}$ transition lies at 440.0 nm--well outside of the
laser bandwidth. Consistent with this interpretation, the intensity of the
SiF 1+1+1 REMPI signals were dependent upon the laser bandwidth. Dulcey and
Hudgens produced the largest SiF ion signals when the laser oscillator was
blocked and broad-band, amplified stimulated emission centered at 440 nm
irradiated the radicals.

B. TRIATOMIC RADICALS

1. CCO Radical

Tjossem and Cool[18] have reported experiments in which they focused a
tunable dye laser within a low pressure hydrocarbon flame and collected the
photoelectrons produced by REMPI of resonant species. They attributed a
REMPI spectrum observed between 443-453 nm to the CCO radical. A portion of
this spectrum is shown in Figure 18.

Experiments with $CH_4/O_2/Ar$, $C_2H_4/O_2/Ar$ and C_2H_6/Ar showed that the
spectrum was most pronounced under lean flame conditions when the laser was
focused somewhat downstream of the visible violet flame zone. It was in this
zone that the concentrations of CCO were expected to be highest.

The CCO spectral carrier was identified using chemical arguments.
Because the spectrum did not change when CD_3CDO and $(CD_3)_2CO$ were injected
into a $D_2/O_2/Ar$ flame, they concluded that the spectral carrier contained no
hydrogen atoms. The strongest spectra were observed when C_3O_2 was injected

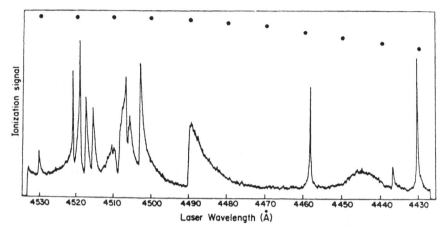

Laser Wavelength (Å)

Reprinted with permission from the 20th Symposium (Intl) on Combustion (The Combustion Institute, Pittsburg, 1984), p.1321

Figure 18. The 3+1 REMPI spectrum attributed to CCO radical between 4430-4530 Å for a $CH_4/O_2/Ar$ (1/2.6/0.1) $\emptyset=0.8$ flame at a pressure of 31 torr. The laser focus was located 8 mm above the burner surface and 1 mm from the ion probe anode surface. Relative laser power (0.8 mJ @ 4500 Å) is plotted by solid points. From Ref. 18.

into an H_2/O_2 flame at 21 torr. C_3O_2 produces CCO radicals by the sequence:[96-98]

$$O + C_3O_2 \quad ----> \quad CO_2 + C_2O$$

$$H + C_3O_2 \quad ----> \quad HC_2O + CO$$

$$R + HC_2O \quad ----> \quad C_2O + RH$$

Under flame conditions energy transfer and pyrolysis may also generate C_2O from C_3O_2.

At the lowest laser intensities (< 2 GW-cm^{-2}) Tjossem and Cool[18] observed that the REMPI signal intensity increased in proportion to the third power of the laser energy. They concluded that the upper electronic state that accounts for the spectrum resides at the sum of three laser photons ($66,225$-$67,700$ cm^{-1}). Ionization occurs through a 3+1 mechanism.

2. NH_2 Radical

Glownia et al.[3] have reported REMPI spectra from NH_2 radicals observed during three photon spectroscopic studies of NH_3. In these studies NH_3 was expanded from a nozzle at a pressure of ~1 atm into a 5×10^{-5} torr vacuum. A laser beam focused onto the expanded gas ionized the resonant molecules. An ion-electron multiplier detected the laser generated ions without mass resolving them. The NH_2 ($\tilde{X}$ $^2B_1$) radicals were produced by predissociation of NH_3 $\tilde{A}$ $^2A_2''$ ($3sa_1'$) molecules. NH_3 $\tilde{A}$ $^2A_2''$ ($3sa_1'$) molecules were prepared by three photon absorption.

The NH_2 radicals were identified by their correspondence with known transitions. The REMPI spectrum of NH_2 $\tilde{X}$ $^2B_1$ radicals was observed between laser wavelengths of 416-434 nm. The NH_2 radical spectrum was attributed to resonances with the one photon $\tilde{A}$ $^2A_1$ (v_2'=16,17)$\leftarrow\tilde{X}$ $^2B_1$ (v_2''=0) bands.[99] To ionize, the laser excited NH_2 $\tilde{A}$ $^2A_1$ (v_2'=16,17) radical had to absorb (at least) three more photons, i.e. a 1+3 REMPI mechanism.[3]

In the wavelength interval of the study the predissociation step which formed the NH_2 radicals was in competition with a 3+1 REMPI process which produced the NH_3 $\tilde{A}$ $^2A_2''$ ($3sa_1'$)$\leftrightarrow\tilde{X}$ $^1A_1'$ spectrum. At shorter laser wavelengths the REMPI spectrum showed weaker REMPI signals from NH_3 and stronger REMPI signals from NH_2. Glownia et al.[3] interpreted this observation as evidence that the NH_3 predissociation rate increased with increasing quanta in the NH_3 v_2' umbrella mode.

3. HCO Radical

The formyl radical, HCO, is an important intermediate in atmospheric photochemical reactions and hydrocarbon combustion. The kinetics of reactions involving HCO photofragments have been studied with laser intracavity absorption[100-102] and LIF methods.[103] Tjossem et al.[19] have reported the first REMPI detection of $\tilde{X}\,^2\Pi$ (A') formyl radicals.

These experiments were conducted at 10 torr in a glass ionization cell through which acetaldehyde was slowly flowed. Photolysis of acetaldehyde with a 308 nm laser produced the formyl radicals. The radicals were probed by a tunable dye laser beam that was focused with a 10 cm lens and overlapped with the photolysis laser beam. The photoelectrons generated by the REMPI process was detected by a platinum bead anode probe. Under these conditions the single laser pulse REMPI detection limit (S/N=2) for HCO is 10^{13} radicals-cm^{-3}, an improvement by a factor of 10-100 over LIF detection.

REMPI signals were observed between 373-427 nm. Figure 19 displays the spectra of experiments that proved the identity of the radical carrier. The laser photolysis products of CH_3CDO and CD_3CDO exhibited identical spectra, but the spectrum of the photolysis products of CH_3CHO differed in appearance. The photolysis of acetaldehyde proceeds via:

$$CH_3CDO \quad ----> \quad CH_3 \;+\; DCO$$
$$CD_3CDO \quad ----> \quad CD_3 \;+\; DCO$$
$$CH_3CHO \quad ----> \quad CH_3 \;+\; HCO$$

Thus, the assignment of the spectral carrier to formyl radical is consistent with the similarity and differences in appearance among the REMPI spectra.

Tjossem et al. observed that the ionization signal varied as the square

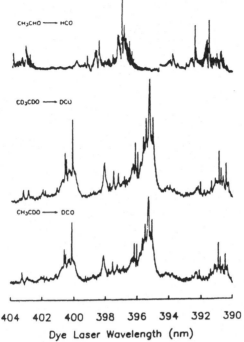

404 402 400 398 396 394 392 390
Dye Laser Wavelength (nm)

Figure 19. The 2+1 REMPI spectra attributed to formyl radical between 390-404 nm. Photolysis with 308 nm laser produced formyl from three isotopically substituted acetaldehydes. Top: CH_3CHO, Middle: CD_3CDO and Bottom: CH_3CDO. From Ref. 19.

of the laser pulse energy. They interpreted this power dependency as evidence that the electronic state which generated the spectrum resided at the sum of two laser photons. Since the ionization potential of HCO is 8.55(±0.01) eV,[104] absorption of one more laser photon could ionize the HCO radicals, i.e. a 2+1 REMPI mechanism.

The spectrum of HCO shows a long series of sub-bands between 373-427 nm (48,117-53,086 cm^{-1}). The progressions are indicative of the bent-linear geometry change in HCO between the ground and excited states. In the earliest paper these series were organized into vibrational progressions from two electronic origins tentatively assigned to the 3p $^2\Sigma^+$ and 3p $^2\Pi$ Rydberg states. Further analysis now shows that both progressions originate from 3p $^2\Pi \leftrightarrow \tilde{X}\ ^2\Pi$ transitions of the type $(0v_2'0) \leftrightarrow (000)$ and $(0v_2'1) \leftrightarrow (000)$.[105] Table 11 summaries the laser wavelengths and vibrational assignments of the subbands. In Table 11 the quantum number K= $|\Lambda \pm \ell_2|$ where ℓ_2 is the vibrational angular momentum associated with the bending mode ($\ell_2 = v_2$, $v_2 - 2, \ldots,$ 1 or 0). The selection rule is $\Delta K = 0, \pm 1, \pm 2$, but in the HCO/DCO 3p $^2\Pi$ Rydberg spectra the $\Delta K = \pm 2$ bands are strongest.

A second series of subbands appeared between 418-432 nm that are tentatively assigned to 3s $^2\Sigma^+$ Rydberg states (Table 12). Numerous unidentified vibrational bands also listed in Table 12 appeared in these REMPI spectra.

Table 13 lists the vibrational frequencies derived from the REMPI spectrum of the 3p $^2\Pi$ state. The ground state frequencies of the formyl radical and cation are also listed. The frequencies observed of the cation and frequencies derived from this new REMPI band system are essentially the same. These resemblances strongly support the assignment of the REMPI spectrum to a Rydberg state.

Many members of the vibrational progressions in the 3p $^2\Pi$ spectrum showed recognizable O, P, Q, R, S branches. Tjossem et al. have analyzed several of these bands by contour simulations, e.g. Figure 20. In their simulations they used the Placzek-Teller rotational line strengths and a rotational temperature of 300K. The simulations found that the rotational constants of the linear 3p $^2\Pi$ upper states lie between B'=1.493-1.498 cm^{-1}.

Table 11. Subbands observed in the spectrum of HCO ($3p\ ^2\Pi \leftrightarrow \tilde{X}\ ^2\Pi$).

v_2'	$(0v_2'0)$		$(0v_2'1)$	
	λ_{LASER} (nm)	$2h\nu$ (cm^{-1})	λ_{LASER} (nm)	$2h\nu$ (cm^{-1})
$(K',K'') = (1,3)$				
10	375.42	53 258.5		
8	386.50	51 732.0		
6	398.45	50 180.9	381.97	52 345.4
4	411.38	48 604.0	393.80	50 773.5
2			406.56	49 180.0
0			420.47	47 554.4
$(K',K'') = (1,2)$				
10	374.63	53 371.1		
8	385.66	51 844.5		
6	397.57	50 292.4	381.16	52 457.5
4	410.45	48 714.7	392.94	50 885.0
2			405.72	49 282(4)
$(K',K'') = (1,1)$				
10	374.17	53 437.1		
8	385.17	51 911.3		
6	397.03	50 360.3	380.67	52 524.8
4	409.88	48 782.7	392.41	50 953.4
2	423.84	47 176.0	405.14	49 352.5
0			418.95	47 726.6
$(K',K'') = (0,2)$				
9	380.07	52 607.6		
7	391.52	51 068.8	375.62	53 231.0
5	403.89	49 505.6	386.94	51 673.7
3	417.40	47 914.8	399.19	50 088.7
$(K',K'') = (2,0)$				
9	379.34	52 708.5		
7	390.76	51 169.1	374.92	53 329.9
5	403.09	49 603.7	386.21	51 771.8
3	416.50	48 007.5		

<u>Table 12.</u> Other 2+1 REMPI bands observed in the spectrum of HCO radical.

λ_{LASER} (nm)	$2h\nu$ (cm^{-1})	λ_{LASER} (nm)	$2h\nu$ (cm^{-1})
3s $^2\Sigma^+$ state progression		<u>Other bands</u>	
418.44	47 785	432.47	46 235
418.99	47 722	406.40	49 200
424.57	47 095	439.34	45 512
425.28	47 016	425.28	47 016
426.20	46 915	412.09	48 521

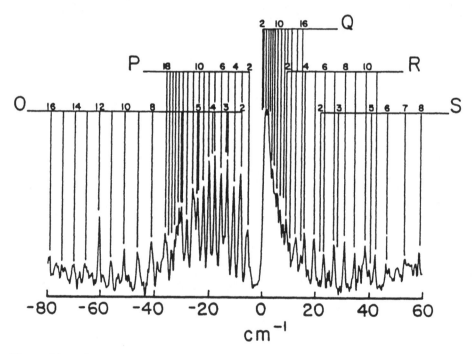

<u>Figure 20.</u> The rotationally resolved 2+1 REMPI spectrum of the (K',K")=(0,2) subband of the 3p $^2\Pi$(070)↔↔X̄ $^2\Pi$(000) transition. From Ref. 19.

Table 13. Comparison of the fundamental frequencies of the formyl radical and cation.

Mode	Ground State (cm^{-1})	3p $^2\Pi$ Rydberg State (cm^{-1})[a]	Cation (cm^{-1})
HCO			
ν_1 CH stretch	2 440(30)[b]		3 088.74[c]
ν_2 Bend	1 080.76[d]	822.1(7)	829.72[e]
ν_3 CO stretch	1 868.17[f]	2 177(3)	2 183.95[g]
DCO			
ν_1 DCO stretch	1 909.77[f]		2 584.56[h]
ν_2 Bend	847.4[i]	657(2)	655(25)[j]
ν_3 CO stretch	1 794.59[f]	1 900(5)	1 904.15[k]

[a] Ref. 105.
[b] Ref. 103 and 106.
[c] Ref. 107.
[d] Ref. 108-110.
[e] Ref. 111.
[f] Ref. 112.

[g] Ref. 113.
[h] Ref. 114.
[i] Ref. 108.
[j] Ref. 115.
[k] Ref. 116.

4. SiF_2 Radical

The SiF_2 radical is a conspicuous product observed during the etching of silicon surfaces of semiconductors. Its electronic spectrum has been previously characterized by UV absorption[117] and LIF spectroscopy.[118,119] Dagata et al.[20] have recently reported REMPI ionization of SiF_2 between 320-322 nm.

SiF_2 was produced in a 800K reactor by the reaction of F_2 and NF_3 with solid silicon. The REMPI spectrum was carried only by m/z 66 (SiF_2^+). Dagata

et al.[20] assigned the REMPI mechanism to two photon resonances with the
$\tilde{B}\ ^1B_2 \leftrightarrow \tilde{X}\ ^1A_1$ transition based upon the coincidence of the REMPI spectrum with
a previous one photon UV study by Gole et al.[117] Between 320-322 nm the
principal features of this incompletely resolved spectrum consisted of a
(0,0) band, a short series of v_2 F-Si-F bend sequence bands, and rotational
structure associated with the 0_0^0 band. The ionization potential of SiF_2 is
10.78(5) eV.[121] A 2+1 ionization mechanism through autoionizing levels can
account for the spectrum.

C. METHYL RADICAL

The methyl radical plays important roles in many reactive systems,
including the early stages of hydrocarbon combustion and atmospheric
chemistry. It serves as a benchmark in theoretical calculations. Until the
advent of REMPI detection schemes, laser spectroscopy could not sensitively
detect methyl radical. Its absorption spectrum lies at wavelengths shorter
than conveniently produced by commercial tunable dye lasers. In addition,
methyl radical has no fluorescence spectrum.

Overwhelming experimental[122,123] and theoretical evidence[124,125] has
established that methyl radicals are of planar D_{3h} symmetry with an $\tilde{X}\ ^2A_2''$
ground state symmetry. The radical electron resides within the nonbonding $2a_2''$
orbital which is mostly composed of the carbon atom p_z orbital.

The one photon UV and VUV absorption spectra of methyl radicals were
exhaustively studied by Herzberg and Shoosmith.[126,127] Three Rydberg series;
ns $^2A_1'$, nd $^2E''$, and nd $^2A_1'$; were observed and analyzed. Other Rydberg series
were not detected because the selection rules under D_{3h} symmetry formally
"forbid" one photon transitions to these states. In fact an examination of

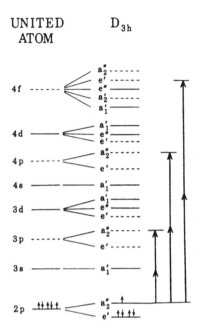

UNITED D_{3h}
ATOM

<u>Figure 21.</u> The correlation diagram between the united atom, fluorine, and the D_{3h} symmetry orbitals of methyl radical. States accessible by one photon absorption from the 2p $^2A_2''$ ground state are indicated by solid lines. Dashed lines indicate states accessible via multiphoton transitions. Arrows indicate the two photon transitions to states observed in Refs. 9 and 10. This diagram does not show the $^2E'$ valence state of methyl radical. From Ref. 10.

Reprinted with permission from J. Chem. Phys. 79, 571 (1983) Copyright (1983) American Institute of Physics

the symmetries of the higher electronic states of methyl radical (Figure 21) shows that more than half of the electronic states can only be reached via multiphoton absorption!

Table 14 lists the D_{3h} electronic state symmetries and shows the selection rules for transitions from the methyl radical $\tilde{X}$ $^2A_2''$ state in experiments which use only one laser frequency. In principle, the assemblage

Table 14. Optical selection rules for transitions from the vibrationless $\tilde{X}$ $^2A_2''$ state of methyl radical to singly excited states of other symmetries.[a]

Final state symmetry species $(\Phi_e\Phi_v)$	Number of Simultaneously Absorbed Identical Photons		
	One Photon	Two Photon	Three Photon
$^2A_1'$	$\Delta K=0$ $K=0$; $\Delta N=\pm1$ $K\neq0$; $\Delta N=0, \pm1$	f[b]	$\Delta K=0$ $K=0$; $\Delta N=\pm1, \pm3$ $K\neq0$; $\Delta N=0, \pm1, \pm2, \pm3$
$^2A_1''$	f	$\Delta K=0$ $K=0$; $\Delta N=\pm1$ $K\neq0$; $\Delta N=0, \pm1, \pm2$	$\Delta K=\pm3$ $K''=0$; $\Delta N=\pm1, \pm3$ $K''\neq0$; $\Delta N=0, \pm1, \pm2, \pm3$
$^2A_2'$	f	f	f
$^2A_2''$	f	$\Delta K=0$ $K=0$; $\Delta N=0, \pm2$ $K\neq0$; $\Delta N=0, \pm1, \pm2$	$\Delta K=\pm3$ $K''=0$; $\Delta N=0, \pm2$ $K''\neq0$; $\Delta N=0, \pm1, \pm2, \pm3$
$^2E'$	f	$\Delta K=\Delta\ell=\pm1$ $\Delta N=0, \pm1, \pm2$	$\Delta K=2(-\Delta\ell)=\pm2$ $\Delta N=0, \pm1, \pm2, \pm3$
$^2E''$	$\Delta K=\Delta\ell=\pm1$ $\Delta N=0, \pm1$	$\Delta=2(-\Delta\ell)=\pm2$ $\Delta N=0, \pm1, \pm2$	$\Delta K=\Delta\ell=\pm1$ $\Delta N=0, \pm1, \pm2, \pm3$

[a] Hund's case (b) notation.   [b] f = forbidden process.

of one and two photon absorption experiments can observe all of the s, p, and d Rydberg states and the $^2E'$ valence state. However, transitions of the type, $^2A_2' \leftrightarrow {}^2A_2''$, are forbidden in one frequency experiments, but may occur when the radicals are irradiated with two different laser frequencies.[39] States of $^2A_1'$ symmetry are unimportant for the spectroscopy of the lower Rydberg states of methyl radical.

In the following three sections REMPI spectroscopic studies of methyl radical which have used one, two, and three photon absorption to prepare the resonant states are described. The one and three photon resonant REMPI experiments were based upon UV/VUV absorption data previously reported by Herzberg.[126,127] In the third section the new electronic states discovered by two photon resonant REMPI spectroscopy are discussed.

1. One Photon Resonance Enhanced Ionization.

Danon et al.[8] demonstrated laser ionization of methyl radicals by a 1+1 ionization mechanism. In this experiment a tunable 215 nm laser beam excited $\tilde{X}\ ^2A_2''$ methyl radicals into the 3s $^2A_1'$ Rydberg state. When the 3s $^2A_1'$ radicals absorbed one photon from a second laser tuned to 266 nm, methyl cations were produced. This ionization technique required that the two laser beams overlap in space and that the two light pulses overlap in time. The tunable 215 nm laser light (30 μJ/pulse; 5 nsec width) was produced by Raman shifting a 294 nm laser beam in high pressure H_2 gas. The 266 nm light (10 mJ/pulse; 7 nsec width) was produced from a frequency quadrupled Nd:YAG laser. During this study methyl radicals were generated by 266 nm photolysis from methyl iodide.

The rotational lines in the REMPI spectra of CH_3 and CD_3 appeared broadened. Figure 22 shows the 1+1 REMPI spectrum of CD_3. Based upon the widths of individual rotational lines, Danon et al.[8] estimated that the lifetimes of the 3s $^2A_1'$ state in CH_3 and CD_3 are 0.12 and 1.2 psec respectively.

The 1+1 REMPI excitation scheme presented by Danon et al. seemed very sensitive. Assuming that the laser beams irradiated 2×10^4 radicals, they estimated a detectivity for methyl radicals of 3×10^7 radicals-cm^{-3}.

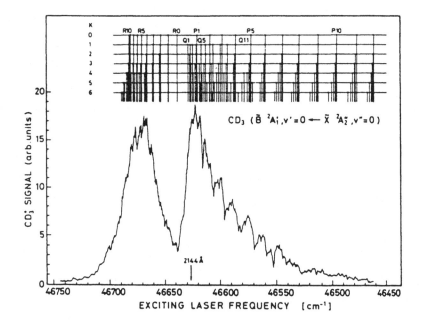

Figure 22. The two laser frequency 1+1 REMPI spectrum of CD_3 between 214-217 nm showing the $^2A_1' \leftarrow \tilde{X}\ ^2A_2''$ band. From Ref. 8.

Reprinted with permission from J. Chem. Phys. 76, 2399 (1982) Copyright (1982) American Institute of Physics

2. Three Photon REMPI of Methyl radicals.

DiGiuseppe et al.[5,6] reported the first REMPI spectrum of methyl radicals. In their studies the nd $^2E''$, 4d $^2A_1'$, and 4s $^2A_1'$ states previously observed with one photon VUV spectroscopy were prepared by simultaneous three photon absorption between 410-460 nm. Absorption of one more photon ionized the excited state radicals, i.e. a 3+1 mechanism.

The methyl radicals were produced by pyrolysis of azomethane, methyl iodide, di-t-butyl peroxide and dimethyl sulfoxide in a tantalum foil oven.

The assignment of the m/z 15 spectral carrier to methyl radical was confirmed by the identical appearance of the REMPI spectra observed in the pyrolysis products of these precursors and by dependence of the m/z 15 signal intensity upon the oven temperature. Very intense laser conditions were used to generate these spectra. The laser used for these experiments produced a 12 mJ, 0.5 cm^{-1} FWHM bandwidth, 10 nsec pulse which was focused into the ion optics by a 50 mm focal length lens.

Figures 23 and 24 show the REMPI spectrum of CH_3 between 414-453 nm. The spectra display 4s $^2A_1'$, 3d $^2A_1'$, 3d $^2E''$, and 4d $^2E''$ origins and vibrational structure from the out-of-plane bend, ν_2. Table 15 lists the REMPI bands observed from CH_3 and CD_3. As expected of nontotally symmetric vibrational mode, the out-of-plane bending mode displays a progression of bands that follow the selection rule $\Delta v_2 = 0, \pm 2, \ldots$

Table 16 compares the out-of-plane bending frequencies of the ground and Rydberg states. Consistent with the view that the Rydberg electron contributes little to the Rydberg core bonding, the Rydberg and cation[128] out-of-plane bending frequencies are of similar magnitude.[128,129]

3. Two Photon Spectroscopy.

Hudgens et al.[9,10] observed spectra of methyl radicals between 260-336 nm. These states were members of two Rydberg series which they assigned as np $^2A_2''$ and (tentatively) nf $^2E'$ Rydberg series (Table 17). Inspection of the selection rules (Table 14) shows that Rydberg series of these symmetries are two photon "allowed" but one photon "forbidden".

During these studies methyl radicals were produced by pyrolysis of precursors in a tantalum oven and by the reaction of F + CH_4 in a flow reactor.

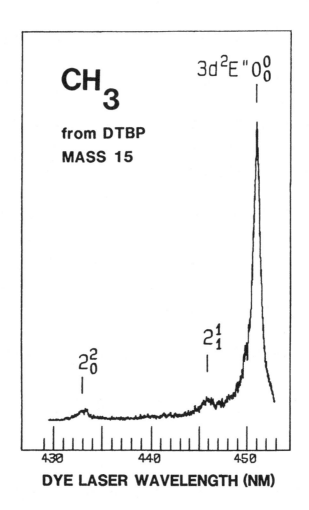

<u>Figure 23.</u> The 3+1 REMPI spectrum of CH_3 radical (m/z 15) between 427-453 nm using pyrolyzed di-t-butylperoxide as the methyl radical source. From Ref. 6.

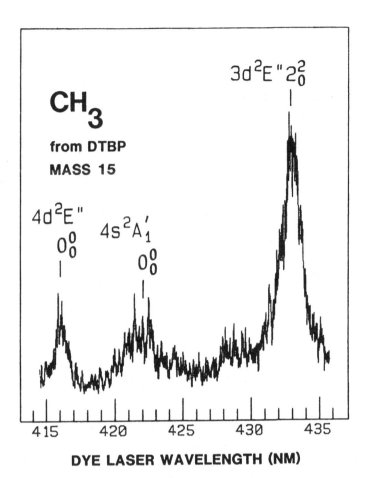

Figure 24. The 3+1 REMPI spectrum of CH_3 radical (m/z 15) between 414-436 nm using pyrolyzed di-t-butylperoxide as the methyl radical source. From Ref. 6.

Table 15. The 3+1 REMPI bands and assignments observed in CH_3 and CD_3. [a]

| | CH$_3$ Radical | | CD$_3$ Radical | | Isotopic Shift |
	λ_{LASER} (nm)	2hν (cm^{-1})	λ_{LASER} (nm)	2hν (cm^{-1})	(cm^{-1})
TRANSITION					
3d ^{2}E" 0_0^0	450.8	66 530	451.3	66 456	-74
2_1^1	445.8	67 276	447.2	67 066	-210
2_0^2	432.9	69 281	437.7	68 521	-760
2_1^3	428.4	70 008	433.4	69 200	-808
4s ^{2}A$_1'$ 0_0^0	(421.9)	71 086	423.4	70 834	-252
2_1^1	---	---	419.0	71 578	
4d ^{2}E" 0_0^0	415.9	72 110	415.8	72 130	+20
4d ^{2}A$_1'$ 0_0^0	---	---	414.9	72 285	

(a) Data are taken from Ref. 6.

Table 16. Comparison of the vibrational frequencies of CH_3 and CD_3 ground
states, Rydberg states, and cations.

		RYDBERG STATES		
Normal Mode	$\tilde{X}\,^2A_2''$ (cm^{-1})	$3p\,^2A_2''$ (cm^{-1})[a]	$3d\,^2E'$ (cm^{-1})[b]	Cation (cm^{-1})
CH_3				
ν_1 CH sym. stretch	3004.8[c]	2914[b]	----	(2903)[d]
ν_2 CH_3 out-of-plane bend	606·453[e]	1334[b]	1372[d]	1380[f]
CD_3				
ν_1 CD sym. stretch	2136[g]	2031[b]	----	----
ν_2 CD_3 out-of-plane bend	453[h]	1032[b]	1031[d]	1070[f]

[a] Ref. 10.
[b] Ref. 6.
[c] Ref. 130.
[d] Predicted by *ab initio* calculations of Ref. 131.
[e] Ref. 122, 132.
[f] Ref. 129.
[g] Ref. 133.
[h] Argon matrix data from Ref. 134.

i. The np $^2A_2'' \leftrightarrow \tilde{X}\,^2A_2''$ Bands

Figures 25-28 show REMPI spectra of CH_3 and CD_3 between 284-336 nm.
Table 17 lists the observed REMPI bands and their assignments in CH_3 and CD_3
respectively. The origin of each electronic state was identified by its
relatively small (≤86 cm^{-1}) isotope shift. The series of bands in CH_3 at

242

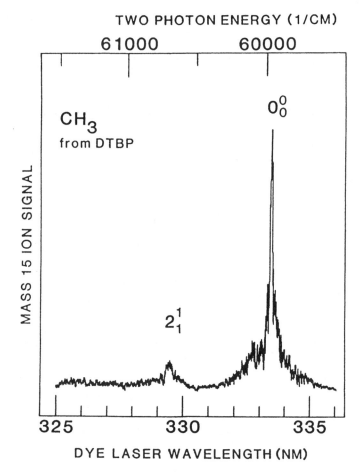

Reprinted with permission from J. Chem. Phys. 79, 571 (1983) Copyright (1983) American Institute of Physics

Figure 25. The 2+1 REMPI spectrum of CH_3 (m/z 15) between 325-336 nm showing the 3p $^2A_2''$ origin and out-of-plane bending mode hot band. CH_3 was produced by pyrolysis of di-t-butyl peroxide. From Ref. 10.

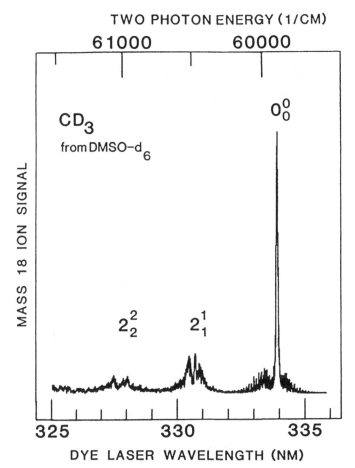

Figure 26. The 2+1 REMPI·spectrum of CD_3 (m/z 18) between 325-336 nm showing the 3p $^2A_2''$ origin and out-of-plane bending mode hot bands. CD_3 was produced by pyrolysis of dimethylsulfoxide-d_6. From Ref. 10.

244

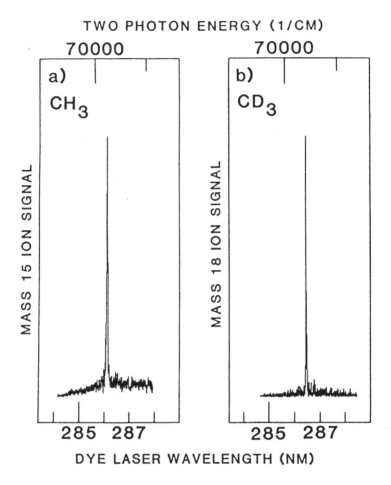

Reprinted with permission from J. Chem. Phys. 79, 571 (1983) Copyright (1983) American Institute of Physics

Figure 27. The 2+1 REMPI spectra of the 4p $^2A_2''$ 0_0^0 bands in a) CH_3 (m/z 15) and b) CD_3 (m/z 18). From Ref. 10.

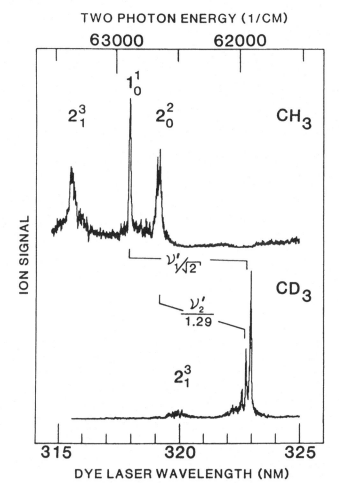

Reprinted with permission from J. Chem. Phys. 79, 571 (1983) Copyright (1983) American Institute of Physics

Figure 28. The 2+1 REMPI spectra of CH_3 (upper trace; m/z 15) and CD_3 (lower trace; m/z 18) between 315-325 nm showing vibrational structure in the 3p $^2A_2''$ Rydberg state. Correlation lines between the two spectra show the expected harmonic potential isotope shifts of the 2_0^2 and the 1_0^1 bands from CH_3 to CD_3. From Ref. 10.

Table 17. The 2+1 REMPI bands and assignments observed in CH_3 and CD_3.[a]

TRANSITION	CH₃ Radical		CD₃ Radical		Isotopic Shift
	λ_{LASER} (nm)	$2h\nu$ (cm⁻¹)	λ_{LASER} (nm)	$2h\nu$ (cm⁻¹)	(cm⁻¹)
3p $^2A_2''$ 2_2^0	340.8[b]	58 692			
0_0^0	333.4	59 972	333.9	59 886	-86
2_1^1	329.4	60 700	330.7	60 466	-234
2_2^2			327.8	60 995	
2_0^2	319.1	62 660	322.75	61 955	-705
1_0^1	317.9	62 886	322.9	61 917	-969
2_1^3	315.6	63 355	320.0	62 473	-882
2_0^4	306.2	65 300	312.7	63 941	-1359
4p $^2A_2''$ 0_0^0	286.3	69 837	286.5	69 789	-48
5p $^2A_2''$ 0_0^0			271.5	73 645	
6p $^2A_2''$ 0_0^0			264.7	75 557	
4f $^2E'$ 0_0^0	275.75	72 508	276.05	72 431	-77
5f $^2E'$ 0_0^0	266.7	74 961	267.0	74 885	-76

(a) Except where noted, data are taken from Ref. 10.

(b) From Ref. 135.

Reprinted with permission from J. Chem. Phys. 79, 571 (1983) Copyright (1983) American Institute of Physics

333.4 and 286.3 nm and in CD_3 at 333.9, 286.5, 271.5, and 264.7 nm fit a two
photon resonant n=3,4,5,6 Rydberg series of quantum defect 0.6. The quantum
defect of 0.6 identified the series as a np Rydberg series.[44-46] The
intensity of each successive Rydberg member decreased rapidly with increasing
"n". No rotational or vibrational structure was observed for n≥4 members
(e.g. Figure 27).

The np Rydberg states of methyl radical are of $^2A_2''$ and $^2E'$ symmetries.
The two selection rules listed in Table 14 show that $^2A_2''$-$^2A_2''$ transitions and
$^2E'$-$^2A_2''$ follow the selection rules ΔK=0 and ΔK=±1 respectively. An
inspection of the rotational energy levels of an oblate symmetric top given
by the equation:

$$F_e(N,K) = T_e + B_o[N(N+1) - K^2] + D_N N^2(N+1)^2 + C_o(K-\zeta_e \ell)^2$$

shows that even a poorly resolved 2+1 REMPI spectrum can establish the upper
state symmetry. (When the radical is a prolate symmetric top, much higher
resolution data are required.)

In the rotational energy equation B_o and C_o are the commonly used
oblate symmetric top rotational constants ($B_o=2C_o$ for CH_3) and D_N is the
centrifugal stretching term. The term $(K-\zeta_e \ell)$ is the magnitude of the nuclear
top angular momentum, K is the signed quantum number of total angular
momentum about the symmetry axis (K=-N, -N+1,...,+N-1,+N). The Coriolis
term, ζ_e, is the expectation value of the projection of the Rydberg's
electronic angular momentum upon the symmetry axis. Approximate values for
ζ_e are found by using the hydrogenic wavefunction of the Rydberg electron:

$$\zeta_e = <\Phi_{Ryd}| L_z |\Phi_{Ryd}>.$$

In the selection rules shown in Table 14 the term, ℓ, is a bookkeeping variable for the selection rules affecting the Coriolis term used in the calculation of $F_e(N,K)$. For np $^2E'$ and nf $^2E'$ states ζ_e lies between 0.9-1; for the nd $^2E'$ state ζ_e lies near -2; ζ_e is defined to be zero for nondegenerate symmetries.

The 3p 0_0^0 band of CD_3 at 333.9 nm shows resolved rotational structure. Hudgens et al. successfully analyzed the 333.9 nm band of CD_3 and confirmed that this band arose from 3p $^2A_2'' \leftrightarrow \tilde{X}\ ^2A_2''$ transitions. Figure 29 shows the rotationally resolved experimental data, the rotational assignments, and a

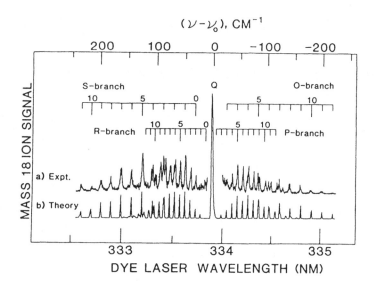

<u>Figure 29.</u> Rotational spectrum and assignments of the O, P, Q, R, and S branches of the 3p $^2A_2''$ band of CD_3. (a) The experimentally observed m/z 18 signal produced using circularly polarized light. The intense Q-branch is not shown. (b) Computer generated simulation of the rotational structure (T_r=300K). From Ref. 10.

computer simulation of the data. The identification of O and S branches ($\Delta J=\pm 2$) verifies the two photon nature of the transitions. The strong Q-branch is symptomatic of the $\Delta K=0$ selection rule and the fact that the ground and excited state have very similar rotational constants. The simulation calculated the transition intensities based upon the two photon rotational line strengths for symmetric tops derived by Chen and Yeung.[34] A rotational temperature of 300K gave the best fit of the band. Table 18 lists rotational constants of methyl radicals used in the simulations and those derived from the results of REMPI experiments.

Table 18. Rotational constants of CH_3 and CD_3.

STATE	T_e (cm^{-1})	B_o (cm^{-1})	$D_N \times 10^5$ (cm^{-1})	Reference
CH_3				
$\tilde{X}\ ^2A_2''$	0	9.578	77	122
$4f\ ^2E'$	72 508	10.2(2); ζ_e=0.3		10
CD_3				
$\tilde{X}\ ^2A_2''$	0	4.79(2)	19(12)	(a)
$3s\ ^2A_1'$	46 627.8(5)	4.385(32)	41(21)	(a)
$3p\ ^2A_2''$	59 886.3(5)	4.76(2)	23(10)	(a)
$3d\ ^2A_1'$	66 725.2(5)	5.05(2)	22(10)	(a)
$4f\ ^2E'$	72 431	5.1(1); ζ_e=0.3		10

(a) From a least squares fit of rotational data reported in Ref. 127 and Ref. 10.

The 3p $^2A_2''$ assignment was also verified by polarization ratio measurements. In this experiment the CH_3 Q-branch intensity ratio, I_l/I_c, was measured as the laser beam polarization was changed from linear to circular. The observed decrease in the REMPI signal as the laser beam became circularly polarized ($I_l/I_c=2.7$) was consistent with the assignment of the 333.9 nm band to 3p $^2A_2''\leftarrow\tilde{X}\,^2A_2''$ transitions. If the 3p 0_0^0 band were of $^2E'$ symmetry, polarization ratio of $I_l/I_c=0.67$ would be expected.[34]

The 3p $^2A_2''$ state showed vibrational activity in the ν_2 out-of-plane bending mode and the ν_1' symmetric stretch in the 3p $^2A_2''$ Rydberg state. The $1\nu_1$ symmetric stretch and $2\nu_2$ levels appeared to mix strongly in CD_3 (Figure 28).

Subsequent work by other research groups has yielded additional data. Smyth and Taylor[135] reported the 2_2^0 hot band of CH_3 observed in a CH_4/air diffusion flame at 340.8 nm. Chen et al.[136] have reported observation of rotational structure in the 3p $^2A_2''$ 0_0^0 band of supersonically cooled CH_3. In their experiments CH_3 radicals were generated by rapid pyrolysis in an oven attached to a pulsed valve. Adiabatic expansion of the pyrolysis products cooled the radicals. Rotational structure up to $N''=4$ was observed.

ii. The nf $^2E'\leftarrow\tilde{X}\,^2A_2''$ Bands

Hudgens et al.[10] identified a second Rydberg series in methyl radical with members near 276 nm and 267 nm (Table 17). The 276 nm bands showed incompletely resolved rotational structure; however, simulations indicated that the upper state was of $^2E'$ symmetry. Two different Rydberg series solutions could account for these bands: a) an f-orbital series of n=4,5 and $\delta=0$, or b) a p-orbital series of n=5,6 and $\delta=1$.

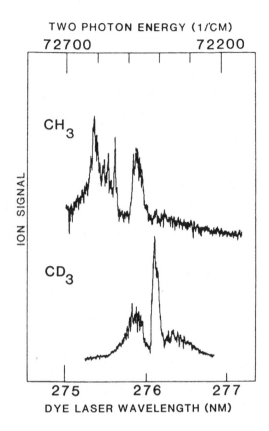

TWO PHOTON ENERGY (1/CM)

ION SIGNAL

CH_3

CD_3

DYE LASER WAVELENGTH (NM)

Figure 30. The 2+1 REMPI spectrum of CH_3 (m/z 15) and CD_3 (m/z 18) between 275-277 nm of the tentatively assigned 3p $^2E' \leftrightarrow \tilde{X}$ $^2A_2''$ bands.

Significant difficulties surrounded assignment of these REMPI bands to the 5p $^2E'$ and 6p $^2E'$ Rydberg states. The quantum defect of $\delta=1$ was anomalously high for a p-orbital Rydberg series. A p-orbital assignment also

left the 3p and 4p Rydberg bands unaccounted for. These bands should have
appeared with greater intensity than the proposed 5p and 6p bands. Because
of these difficulties, the p-orbital Rydberg state assignment was abandoned.

The experimental evidence[10] supported their assignment to the 4f $^2E'$ and
5f $^2E'$ series. The quantum defect of $\delta=0$ is consistent with this assignment.
No expected bands were absent because nf Rydberg series begin with n=4.
Thus, the nf $^2E'$ assignment was offered.

Hudgens et al.[10] noted that the rotational constants required to
simulate the 4f $^2E' \leftarrow \tilde{X}\ ^2A_2''$ spectra were greater than observed in the ground
state. This indicates that the 4fe' Rydberg orbital has a bonding
interaction with the hydrogens. Another indication of this bonding
interaction was the electronic Coriolis value of $\zeta_e=0.3$.

D. SUBSTITUTED METHYL RADICALS.

The vibrational structure observed in REMPI spectra of methyl radical
and of substituted methyl radicals (CH_2F, CH_2OH, $CHCl_2$, and CF_3) reflects the
interactions which involve the unpaired radical electron. These interactions
change between the ground and Rydberg states. All CH_nX_m (X=F,Cl, OH; m+n=3)
radicals examined by REMPI spectroscopy have exhibited vibrational bands
which originate from the out-of-plane (umbrella) bending mode. These bands
stem from the large changes in the out-of-plane bend potential surface which
accompanies the promotion of the unpaired electron into a Rydberg state.

In the ground state the unpaired electron resides mostly on the carbon
atom. As the radical undergoes out-of-plane bending excursions, the unpaired
electron supports rehybridization of carbon orbitals from sp^2 to sp^3. This

rehybridization broadens the restoring potential surface along the out-of-plane bending coordinate. As a result CH_nH_m radicals have low ground state vibrational frequencies.

When the unpaired electron is promoted into a Rydberg state, it can no longer contribute to the rehybridization process. Thus, the out-of-plane restoring potential narrows and the out-of-plane bending mode increases in frequency. Consistent with this view, the known out-of-plane vibrational frequencies of CH_nX_m cations and Rydberg radicals are 100-1,000 cm^{-1} larger than observed for the ground state radicals.

The dissimilarity in the unpaired electron interactions of methyl and substituted methyl radicals is seen in the vibrational structure originating from the other normal modes. In methyl radical the CH symmetric stretching frequency decreases only slightly between the ground and 3p $^2A_2''$ Rydberg state (Table 16). Overall, the unpaired electron in planar methyl radical is relatively isolated within the singly occupied carbon p_z orbital. When the p_z electron is promoted into a Rydberg state or lost through ionization, the C-H bond strength decreases only slightly. This absence of significant change is also seen in the small change in the rotational constants (Table 18).

In contrast, in substituted methyl radicals the p_z-orbital on the O, Cl, and F atoms interacts with the carbon p_z orbital to form π and π^*-orbitals. The unpaired radical electron resides in the π^*-antibonding orbital. When this π^* electron is promoted into a Rydberg orbital or removed by ionization, the carbon-substituent bonding greatly increases. Thus, the Rydberg states of substituted methyl radicals exhibit larger stretching and in-plane bending vibrational frequencies than observed in their ground states. In the following sections the spectral data which led to these conclusions are shown.

1. CH$_2$F Radical

Using REMPI spectroscopy Hudgens et al.[21] observed the first spectrum of any electronic state of the fluoromethyl radical. CH$_2$F and CD$_2$F were produced in a flow reactor from the reactions of F + ketene, F + ketene-d$_2$, and F + CH$_3$F. This study was also the first to identify fluoromethyl radicals as a major product of the F + ketene elementary reaction. The results of mass spectrometry studies of this reaction were discussed in Section III-B.1.

Figures 31-34 show the mass resolved composite spectra of the CH$_2$F and CD$_2$F radicals between 292-382 nm. These spectra display extensive vibrational progressions which terminate at prominent origins in CH$_2$F at 378.25 nm and 316.00 nm and in CD$_2$F at 378.80 and 316.40 nm. Based upon an adiabatic ionization potential of 9.04(0.1) eV,[137] Hudgens et al. fit these origins to a two photon resonant 3p and 4p Rydberg series with a quantum defect of δ~0.6.

The REMPI spectra of the 3p and 4p Rydberg states of CH$_2$F and CD$_2$F display nearly 115 bands. Table 19 lists the positions of band origins and the more significant spectral features. The majority of the REMPI bands were assigned to combination and overtone bands that originated from the activity in four vibrational modes of the Rydberg state and from the activity of the ground state out-of-plane bend.

A normal mode calculation based upon the Wilson FG matrix method[138] was undertaken and found to support the vibrational assignments. These calculations fit the observed vibrational frequencies and enabled a determination of force constants. The calculation revealed strong mixing of the CF stretching and CH$_2$ scissors modes in CH$_2$F. The vibrational analysis also facilitated assignments of the 5p 0_0^0 bands.

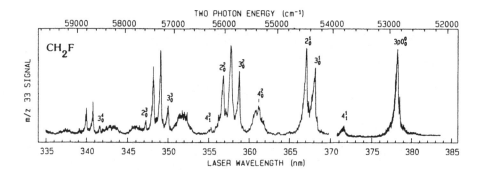

Figure 31. The 2+1 REMPI spectrum of a 3p Rydberg state of CH_2F (m/z 33) between 335-389 nm. From Ref. 21.

Reprinted with permission from J. Chem. Phys. in press, manuscript no. #A7.03.121.

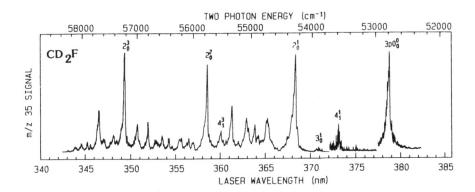

Figure 32. The 2+1 REMPI spectrum of a 3p Rydberg state of CD_2F (m/z 35) between 342-389 nm. From Ref. 21.

Reprinted with permission from J. Chem. Phys. in press, manuscript no. #A7.03.121.

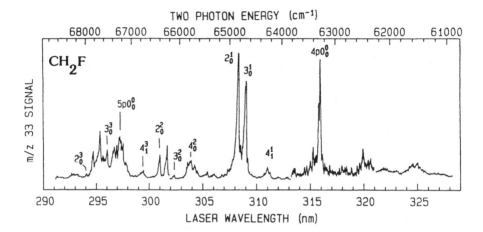

Figure 33. The 2+1 REMPI spectrum of 4p and 5p Rydberg states of CH_2F (m/z 33) between 292-320 nm. From Ref. 21.

Reprinted with permission from J. Chem. Phys. in press, manuscript no. #A7.03.121.

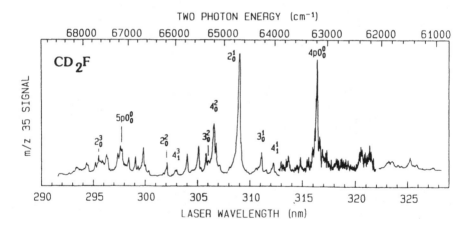

Figure 34. The 2+1 REMPI spectrum of 4p and 5p Rydberg states of CD_2F (m/z 35) between 292-320 nm. From Ref. 21.

Reprinted with permission from J. Chem. Phys. in press, manuscript no. #A7.03.121.

Table 19. The more prominent vibrational bands observed in the 2+1 REMPI spectrum of fluoromethyl radicals.

TRANSITION	CH$_2$F Radical		CD$_2$F Radical	
	λ_{LASER} (nm)	2hν (cm^{-1})	λ_{LASER} (nm)	2hν (cm^{-1})
3p 0_0^0	378.25	52 863	378.80	52 786
4_1^1	371.60	53 808	373.10	53 592
3_0^1	368.35	54 283	371.20	53 866
2_0^1	367.30	54 438	368.30	54 290
1_0^1	---	---	363.80	54 962
4_0^2	361.50	55 311	365.20	54 751
3_0^2	359.05	55 688	364.15	54 909
$2_0^1 3_0^1$	358.00	55 852	361.30	55 342
2_0^2	357.05	56 000	358.50	55 774
4_1^3	355.45	56 252	360.05	55 534
4p 0_0^0	316.00	63 275	316.40	63 195
4_1^1	311.00	64 292	312.25	64 035
3_0^1	308.95	64 719	311.10	64 272
2_0^1	308.30	64 855	309.00	64 708
1_0^1	---	---	305.80	65 385
4_0^2	303.90	65 794	306.60	65 215
3_0^2	302.40	66 120	306.10	65 321
2_0^2	301.00	66 427	302.05	66 197
4_1^3	299.45	66 771	302.90	66 011
5p 0_0^0	297.25	67 265	297.60	67 186

Table 20. Summary of experimentally observed fluoromethyl radical
vibrational frequencies.

Normal Mode	Ground State (cm^{-1})	3p Rydberg State (cm^{-1})[a]	4p Rydberg State (cm^{-1})[a]
CH$_2$F			
ν_2 (CF stretch)	1170.4[b]	1575	1580
ν_3 (CH$_2$ scissors)	----	1420	1443
ν_4 (out-of-plane bend)	260[a]	1223	1259
CD$_2$F			
ν_1 (CH sym. stretch)	----	2176	2190
ν_2 (CF stretch)	1191[c]	1504	1513
ν_3 (CD$_2$ scissors)	1013[d]	1080	1076
ν_4 (out-of-plane bend)	170[a]	976	1004

[a] Ref. 21.

[b] Ref. 139.

[c] Ref. 140, Ar matrix.

[d] Ref. 141, Ar matrix.

Reprinted with permission from J. Chem. Phys. in press, manuscript no. #A7.03.121.

Table 20 lists the active vibrational frequencies and compares them with their ground state frequencies. The $\pi^* \to R$ nature of these electronic transitions is reflected in the 300-400 cm^{-1} larger ν_2' CF stretching modes observed in the Rydberg states.

Table 20 shows that the ν_4 out-of-plane bending mode increases in frequency by a factor of five between the ground and Rydberg states. As discussed above, the small ground state ν_4'' out-of-plane frequency reflects the stabilizing and broadening of the out-of-plane potential surface contributed by rehybridization from sp^2 to sp^3 bonding around the carbon atom.

2. CHCl$_2$ Radical

CHCl$_2$ has been proposed as an intermediate in a number of reactions including the pyrolysis of pentachloroethane[142] and the photolysis of chloroform.[143] Electron spin resonance (ESR)[144], photoelectron spectra[145] and molecular orbital calculations[146] indicate that the radical is essentially planar. A planar cation structure is also expected. The REMPI spectrum reported by Long and Hudgens[24] provided the first analysis of any excited electronic state.

During their study Long and Hudgens[24] generated dichloromethyl radicals by the abstraction of hydrogen from dichloromethane with atomic fluorine. This reaction produces the dichloromethyl radical as its principal product.[147]

Between 355-372 nm REMPI spectra were observed in CHCl$_2$ at m/z 83, 85, and 87 and in CDCl$_2$ at m/z 84, 86, and 88. Since these masses correspond to those expected of the dichloromethyl radical, the assignment of the spectral carrier radical was proven. No REMPI spectrum was observed at masses other than those of the molecular ion. The absence of daughter ions shows that dichloromethyl radicals do not undergo fragmentation subsequent to laser ionization between 355-375 nm.

The REMPI spectra of $CHCl_2$ between 355-375 nm are shown in Figure 35. Table 21 lists the vibrational assignments. The vibrational analysis showed that the REMPI spectrum originated from a Rydberg state lying at the sum of

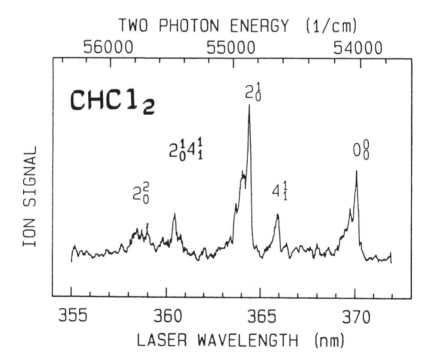

Figure 35. The 2+1 REMPI spectrum of $CHCl_2$ between 355-375 nm. From Ref. 24.

Reprinted with permission from J. Phys. Chem., accepted and scheduled for publication in Sep 87, manuscript #JP8703458-0-6-48

Table 21. The 2+1 REMPI Bands of $CHCl_2$ and $CDCl_2$.[24]

LASER WAVELENGTH (nm)	TWO PHOTON ENERGY (cm^{-1})	FREQUENCY INTERVAL (cm^{-1})	SPECTRAL ASSIGNMENT
$CHCl_2$			
370.1	54 024	0	0_0^0
366.0	54 629	605	4_1^1
364.4	54 869	845	2_0^1
360.4	55 479	1 455	$2_0^1 4_1^1$
358.7	55 742	1 718	2_0^2
$CDCl_2$			
370.4	53 980	0	0_0^0
366.8	54 510	530	4_1^1
364.9	54 794	814	2_0^1
361.4	55 325	1 345	$2_0^1 4_1^1$
359.2	55 664	1 684	2_0^2

Reprinted with permission from J. Phys. Chem., accepted and scheduled for publication in Sep 87, manuscript #JP8703458-0-6-48

two laser photons. Additional experimental evidence supported an assignment to a 3d Rydberg state (ν_{00}=54,024 cm^{-1}) that has a quantum defect of 0.1. Since the ionization potential of $CHCl_2$ radical is 8.32 eV,[145] the ion signal arises from a 2+1 REMPI mechanism.

Vibrational structure originated from activity in the ν_2' (a_1) C-Cl stretching mode and the ν_4 (b_1) out-of-plane bending mode. The C-Cl symmetric stretch frequencies in the 3d Rydberg state are 845 cm^{-1} in $CHCl_2$

and 814 cm^{-1} in CDCl$_2$. These values are essentially the same as those observed in the cation (CHCl$_2^+$; 860(30) cm^{-1} CDCl$_2^+$; 790(30) cm^{-1}).[145]

Hot bands associated with the ν_4 out-of-plane bending mode were observed. The ground state frequency of this mode is unknown, but the spectrum shows that in CHCl$_2$ the out-of-plane bending frequency is 605 cm^{-1} greater in the Rydberg state than in the ground state. As discussed above (Section IV-D), this frequency increase is expected.

3. CH$_2$OH Radical

The hydroxymethyl radical, CH$_2$OH, plays important roles in a wide variety of chemical systems including combustion, atmospheric, and interstellar chemistry. In 1983 using REMPI spectroscopy, Dulcey and Hudgens observed the first electronic spectrum of hydroxymethyl radicals.[48] An assignment of the REMPI mechanism and the identity of the excited state was hampered by imperfect data regarding the CH$_2$OH ionization potential. In 1986 contemporaneous papers by Dulcey and Hudgens[22] and by Bomse, Dougal, and Woodin[148] concluded that laser ionization of hydroxymethyl radicals occurred through a 2+1 REMPI mechanism.

At present REMPI spectroscopy is the only optical method which can sensitively detect CH$_2$OH radicals. Jacox reported that CH$_2$OH photodissociates in an Ar matrix at wavelengths more energetic than 280 nm.[149] A recent theoretical calculation by Solgadi and Flament[150] shows that the excited valence state of CH$_2$OH should rapidly isomerize into ground state CH$_3$O radicals. Thus, future development of LIF detection schemes for CH$_2$OH seems unlikely.

Dulcey and Hudgens[22,48] have reported the REMPI spectra of

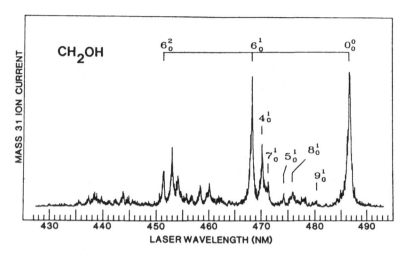

Figure 36. The 2+1 REMPI spectrum of CH_2OH radical (m/z 31) between 425-495 nm. From Ref. 22.
Reprinted with permission from J. Chem. Phys. **84**, 5262 (1986) Copyright (1986) American Institute of Physics

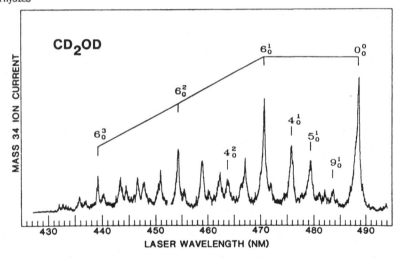

Figure 37. The 2+1 REMPI spectrum of CD_2OD radical (m/z 34) between 425-495 nm. From Ref. 22.
Reprinted with permission from J. Chem. Phys. **84**, 5262 (1986) Copyright (1986) American Institute of Physics

hydroxymethyl radicals, CH_2OH, CH_2OD, CD_2OH, and CD_2OD, produced in a flow reactor by the reaction of chlorine and fluorine atoms with methanol and deuterium substituted methanols. The proof of the spectral carrier was discussed in Section III-B.1. Figures 36-37 show representative REMPI spectra of CH_2OH and CD_2OD. The strong REMPI band near 487 nm in each spectrum was identified as the 0_0^0 band by its small isotopic shift.[22]

The extensive vibrational structure displayed in the spectra indicates that the free radical undergoes significant structural change between the ground and resonant excited states. This structural change is driven by changes in the highest occupied orbitals of the CH_2OH radical which have the configuration: $\ldots(\pi_{CO})^2(n_O)^2(\pi_{CO}^*)^1$. Dulcey and Hudgens[22] offered two different candidate electronic states that could account for the REMPI spectra and stated that these candidate states would possess very different C-O stretching frequencies:

1) If the resonant state was a valence state formed from an $n \to \pi^*$ transition, the CO bond would weaken into a single C-O bond and the REMPI spectrum would display a C-O stretching progression of about ~ 1030 cm^{-1}.

2) If the resonant state was a Rydberg state formed from a $\pi^* \to R$ transition, the CO bond would strengthen into a double bond and the REMPI spectrum would display a C=O stretching mode progression of about ~ 1600 cm^{-1}.

Each spectrum displayed one vibrational progression that possessed an isotopic shift ratio proper for the CO stretching mode. In one photon the CH_2OH CO stretch progression appeared with an interval of $1h\nu \sim 812$ cm^{-1}. This one photon value was too small for a valence C-O stretch, but the two photon value, $2h\nu \sim 1624$ cm^{-1} fell within the range expected of a Rydberg state C=O stretch. Thus, Dulcey and Hudgens assigned the transition to a $\pi^* \to R$

transition (Scheme 2). The calculated quantum defect of this state is $\delta=0.65$ which is of proper magnitude for a 3p Rydberg state. A 2+1 REMPI scheme can exceed the 7.56 eV[151] ionization potential of CH_2OH.

The REMPI spectra of the four isotopic analogues of hydroxymethyl displayed nearly 130 vibrational bands. All important REMPI bands and ~85% of all bands were accounted for by assigning six active vibrational modes in the 3p Rydberg state. A normal mode calculation based upon the Wilson FG matrix method was undertaken and found to support the vibrational assignments. Table 22 compares the experimental vibrational frequencies with the calculated frequencies for the four isotopic analogues. The reasonableness of this vibrational analysis provides further proof of the Rydberg assignment.

Table 23 shows the changes in vibrational frequencies between the ground and Rydberg state hydroxymethyl radicals. Consistent with the $\pi^* \to R$ assignment, the vibrational analysis shows that the hydroxymethyl Rydberg state is more strongly bound than is the ground state radical.

Bomse, Dougal and Woodin[148] reported REMPI spectra of products generated by the infrared multiple photon dissociation (IRMPD) of CH_3OH and CD_3OH. REMPI spectra of CH_2OH and CD_2OH were reported. In these studies an infrared laser was focused into an ionization cell which contained methanol gas at a pressure of 1 torr. After a 0.2-10 μsec delay the IRMPD products were then irradiated by a focused dye laser (3-5 mJ/pulse) tuned between 445-510 nm. Strong REMPI features due to hydroxymethyl radicals were observed. This study showed that the IRMPD process proceeded via:

$$CH_3OH + nh\nu_{IR} \quad ----> \quad CH_2OH + H$$
$$----> \quad CH_3 + OH.$$

Table 22. The vibrational assignments of the normal modes of hydroxymethyl radicals. Calculated frequencies (no parenthesis) based on a Wilson FG matrix fit of the data are shown for each substituted analogue. The observed frequencies are listed in parentheses. From Ref. 22.

Mode	CH_2OH	CH_2OD	CD_2OH	CD_2OD
In-Plane Modes				
ν_1' OH stretch	3452 cm^{-1}	3031 cm^{-1}	3452 cm^{-1}	2526 cm^{-1}
ν_2' CH asym. stretch	3032	2914	2283	2282
ν_3' CH sym. stretch	2914	2525	2150	2147
ν_4' CH$_2$ scissors	1469 (1459)	1446 (1440)	1295 (1298)	1094 (1109)
ν_5' COH bend (mixed)	1109 (1091)	883 (887)	896 (909)	793 (803)
ν_6' CO stretch	1630 (1623)	1619 (1612)	1566 (1568)	1551 (1565)
ν_7' CH$_2$ rock (mixed)	1364 (1351)	1299 (1296)	1071 (1094)	1068
% error in the $\nu_4' - \nu_7'$ block.	0.9%	0.4%	1.0%	1.2%
Out-of-Plane Modes				
ν_8' CH$_2$ wag	956 (950)	953 (949)	774 (787)	766
ν_9' COH torsion	585 (573)	480 (469)	538 (553)	428 (440)
% error in the $\nu_8' + \nu_9'$ block.	1.4%	1.4%	2.2%	2.7%

Table 23. Comparison of vibrational frequencies and normal mode assignments in the ground and 3p Rydberg states of CH_2OH.

Mechanical Action of the Normal Mode	Mode	Ground State Frequency (cm^{-1})[a]	3p Rydberg State Frequency (cm^{-1})[b]	Difference (cm^{-1})
OH stretch	ν_1	3650	3452[c]	-198
CH asym. stretch	ν_2	3019[b]	3032[c]	13
CH symmetric stretch	ν_3	2915[b]	2914[c]	-1
CH_2 scissors	ν_4	1459	1459	0
COH (in plane) bend mixed with CH_2 rock	ν_5	1334	1091	-243
CO stretch	ν_6	1183	1623	440
CH_2 (in-plane) rock mixed with COH bend	ν_7	1048	1351	303
CH_2 (out-of-plane) wag	ν_8	569	950	381
COH torsion	ν_9	420	573	153

[a]Ref. 149.

[b]Ref. 22.

c) The listed value was obtained from the normal mode calculations of Ref. 151.

d) The listed value was obtained from the normal mode calculations Ref. 22.

Based upon polarization measurements, Bomse et al. concluded that the hydroxymethyl radicals were ionized via a 2+1 mechanism through a 3p Rydberg state. For CD_2OH irradiated at 488.0 nm, they reported that the linearly polarized laser beam produced three times more ion signal than observed with the circular polarized laser beam. This polarization ratio is consistent with an assignment to a p-to-p orbital excitation.

Bomse et al. observed significant hot band activity to the red of the 3p 0_0^0 band. This activity was expected because IRMPD generates vibrationally hot products. They assigned this activity to populated ground state ν_4'', ν_5'', ν_6'', ν_7'', ν_8'', ν_9'', and $2\nu_9''$ vibrational levels. For CH_2OH and CD_2OH radicals the maximum internal energy implied by the hot band structure was 1300 ± 40 cm^{-1}.

4. CF$_3$ Radical

The trifluoromethyl radical, CF_3, is ubiquitous in processes such as etching of semiconductor surfaces, halocarbon fire suppression, and ^{13}C-isotope enrichment by infrared laser multiple photon photolysis. ESR studies[155] have shown that the CF_3 radical possesses a pyramidal geometry in its ground electronic state. The CF_3 cation is expected to be planar. This large geometry change upon ionization will cause photoelectron and Rydberg spectra to show long vibrational progressions along the ν_2 out-of-plane bending mode and weak electronic origins. CF_3 radical has no known LIF spectrum.

Figure 38 shows the REMPI spectrum of the CF_3 radical between 415-490 nm reported by Duignan et al.[7,152] They produced CF_3 radicals by pyrolysis of CF_3I in a 1300K tantalum foil oven and by infrared multiple photon dissociation (IRMPD) of CF_3I, CF_3Br, and hexafluoroacetone within the ion

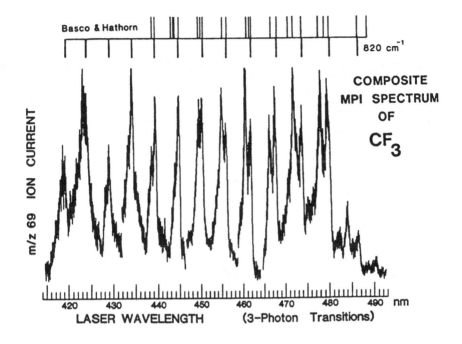

<u>Figure 38.</u> The composite 3+1 REMPI (m/z 69) spectrum of CF_3 radicals
produced by pyrolysis of CF_3I. Uppermost lines denote bandheads reported in
the VUV spectrum of Basco and Hathorn. Lines just below mark 820 cm⁻¹
spacings on either side of the 450.5 nm (66,580 cm⁻¹) bandhead. From Ref. 7.

Reprinted with permission from J. Phys. Chem. **86**, 4156 (1982) Copyright (1982) American Chemical Society

optics of the mass spectrometer. In general the spectra of radicals

generated by IRMPD were broader and less well-resolved than spectra of

radicals generated by pyrolysis. This broadening is probably indicative of

the higher vibrational temperature that photofragments generated by IRMPD

contain as compared to pyrolytic products. A similar effect was noted within

IRMPD generated CH_3 radicals.[152]

Table 24. The bandhead positions observed in the 3+1 REMPI spectrum of CF_3 radical.[7]

λ_{LASER} (nm)	Frequency $(cm^{-1})^a$	λ_{LASER} (nm)	Frequency $(cm^{-1})^a$
486.5	61 640[b]		
484.0	61 970	450.5	66 580[b]
480.2	62 460[b]	449.9	66 660
477.9	62 760	449.6	66 710
474.0	63 270[b]	444.9	67 410[b]
472.1	63 530	443.9	67 570
470.8	63 700	443.3	67 650
467.9	64 100[b]	439.6	68 230[b]
466.4	64 310	438.7	68 360
461.9	64 930[b]	434.3	69 050[b]
461.3	65 020	433.5	69 180
460.7	65 100	429.3	69 860[b]
456.1	65 760[b]	424.4	70 680[b]
455.1	65 900	419.5	71 490[b]

(a) $\pm$ 25 cm^{-1}.   (b) Bandhead in the 820 cm^{-1} progression.

Duignan et al.[7] assigned the laser ionization spectrum to a 3+1 REMPI mechanism. They noted that when simultaneous three photon absorption was adopted for the resonant step, most bandhead positions of the REMPI spectrum agreed closely with the bandhead positions observed in the VUV spectrum between 146-165 nm.[153]   Since the four photon ionization limit is 536 nm (9.25 $\pm$0.4 eV),[154] only one additional laser photon was needed to ionize the highly excited radical.

Between 415-490 nm the REMPI spectrum shows a series of thirteen multiplets. The red-most head of each band is separated by 820 $\pm$10 cm^{-1} (Table 24). The four bands with heads at 419.5, 424.4, 429.3, and 434.3 nm had not been previously attributed to CF_3 radical. These bands also fall in the regular 820 cm^{-1} progression.

Duignan et al. and Basco and Hathorn both believed that the upper state observed by their optical spectra had Rydberg character. This assignment was supported by the fact that the same 820 cm^{-1} vibrational intervals observed in the 3+1 REMPI spectrum and in the VUV study were also seen in the photoionization spectrum of CF_3. In the photoionization study this spacing was interpreted as the ν_2 out-of-plane bending frequency of the ion.[154] Spectroscopic data and molecular orbital calculations that have appeared since this work also support an assignment the REMPI and VUV spectra to the 4p Rydberg states.[156]

Like other substituted methyl radicals, the 820 cm^{-1} ν_2' out-of-plane vibrational frequency observed in the Rydberg state of CF_3 is greater than the ground state frequency. For comparison, the ground state ν_2'' out-of-plane bending vibrational frequency is 701 cm^{-1}.[157,158]

E. OTHER POLYATOMIC RADICALS

1. Allyl and 2-Methylallyl Radicals

The allyl radical ($CH_2 \dot{-} CH \dot{-} CH_2$) is the simplest conjugated hydrocarbon radical. It appears in many thermal[159], photochemical[160,161], and surface catalytic[162,163] reactions. The allyl and 2-methylallyl radicals are bent with C_{2v} symmetry[164-168]. Extensive molecular orbital calculations have focused upon the conjugated π-system; particularly for development of generalized valence bond theory. In the simplest molecular orbital construction the π-orbitals are described by a doubly occupied bonding b_1 orbital, a singly occupied nonbonding a_2 orbital, and an unoccupied antibonding b_1 orbital. The ground state configuration and symmetry is $\tilde{X}\,^2A_2$.

One photon UV absorption studies have reported the lowest energy $^2B_1$ states in allyl radicals produced by flash photolysis. The valence $^2B_1$ state[169] gave rise to bands around 408.3 nm and the Rydberg $^2B_1$ state[170] produced absorption bands between 220-260 nm. No fluorescence of LIF spectra of allyl or 2-methylallyl radicals are known.

Hudgens and Dulcey[26] reported REMPI spectra of allyl radicals between 480-535 nm. During their study the allyl radicals were generated by reactions; F + cyclopropane, F + propene, and Cl + propene; which produced essentially the same m/z 41 ion spectrum. The spectrum was assigned to allyl radical because it is the only thermodynamically allowed product common to these reactions. Consistent with this assignment, the REMPI spectrum observed from the reaction of F + propene-d_6 was carried by m/z 46. Allyl and allyl-d_5 radicals each exhibited one regular vibrational progression which terminated at 498.8 nm and 498.3 nm respectively (Table 25 and 26). These bands were assigned as the electronic origins. Figure 39a shows a REMPI spectrum similar of the one reported by Hudgens and Dulcey.

Figure 40 shows the spectrum of the 2-methylallyl radical between laser wavelengths of 485-543 nm. The same spectrum was observed whenever F atoms or Cl atoms were reacted with isobutene. The spectrum shows vibrational progressions (Table 27) similar to those observed in allyl and allyl-d_5. The spectrum is red shifted by about 22 nm and shows one regular vibrational progression and strong underlying continuous structure. The red-most member of the progression at 521.1 nm was assigned to the origin.

Hudgens and Dulcey found the photon order of the resonant state by reasoning that the electronic configuration of the allyl and 2-methylallyl

Table 25. Spectral features of the allyl radical observed with 2+2 REMPI spectroscopy. From Ref. 25.

Spectral Assignment[a]	λ_{Laser} (nm)	Two Photon Energy, cm^{-1}	Relative Energy, cm^{-1}
A_0^2	489.48	40 848	792
A_1^3	489.82	40 820	764
	490.27	40 782	726
	491.82	40 654	598
	492.64	40 586	530
	493.10	40 548	492
A_0^1	494.30	40 450	394
A_1^2	494.69	40 418	362
A_2^3	495.13	40 382	326
[A_3^4]	495.52	40 350	294
	495.99	40 312	256
	497.94	40 154	98
0_0^0	499.16	40 056.8	0
A_1^1	499.56	40 024	-32
A_2^2	500.04	39 986	-70
[A_3^3]	500.36	39 960	-96
	500.89	39 918	-138
	501.36	39 880	-176
	503.28	39 728	-328
A_1^0	504.53	39 630	-426
A_2^1	505.11	39 584	-472
	505.57	39 548	-508
	506.06	39 510	-546
A_2^0	510.19	39 190	-866
A_3^1	510.79	39 144	-912

[a] "A" refers to the C-C-C (a_1) bending mode.

Table 26. Spectral features of allyl-d_5 radicals observed with REMPI spectroscopy. From Ref. 26.

Spectral Assignment[a]	λ_{Laser} (nm)	Two Photon Energy,[b] cm^{-1}	Relative Energy, cm^{-1}
	488.3	40 947	+822
A_0^2	490.2	40 789	+664
A_0^1	494.1	40 467	+342
0_0^0	498.3	40 125	0
	501.4	39 877	-248
	502.6	39 782	-343
	509.0	39 282	-843
	(510)	39 205	-920
	512.0	39 052	-1 073
	513.3	38 953	-1 172

[a] "A" refers to the C-C-C (a_1) bending mode.   [b] +/-30 cm^{-1}.

Reprinted with permission from J. Phys. Chem. 89, 1505 (1985) Copyright (1985) American Chemical Society

Table 27. Spectral features of the 2-methylallyl radical observed with 2+2 REMPI spectroscopy. From Ref. 26.

Spectral Assignment[a]	λ_{Laser} (nm)	Two Photon Energy,[b] cm^{-1}	Relative Energy, cm^{-1}
A_0^2	509.6	39 235	+866
A_0^1	515.4	38 801	+432
0_0^0	521.1	38 369	0
	526.5	37 976	-394
	527.5	37 904	-466

[a] "A" refers to the C-C-C (a_1) bending mode.   [b] +/-30 cm^{-1}.

Reprinted with permission from J. Phys. Chem. 89, 1505 (1985) Copyright (1985) American Chemical Society

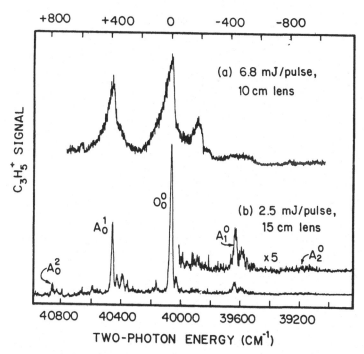

Figure 39. The 2+2 REMPI spectra of cold allyl radical created from 193 nm photolysis of C_3H_5Cl in 4 atm argon. (a) Spectrum observed with 6.6 mJ/pulse of dye laser energy focused by a 100 mm lens (This spectrum is similar to the one reported in Ref. 26); (b) Spectrum observed with 2.5 mJ/pulse of dye laser energy focused with a 150 mm lens. From Ref. 25.

both originated from the same Rydberg state orbital. In this view the quantum defects between the two radicals would be essentially the same and the frequency shift of the band origins between these radicals would reflect the difference in the ionization potentials. The adiabatic ionization potential of the allyl and 2-methylallyl radicals are $IP_a=8.13$ eV[171] and

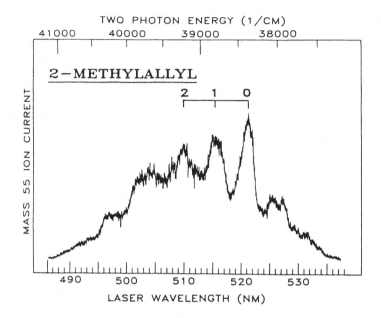

Figure 40. The spectrum of 2-methylallyl radical between 495-538 nm. From Ref. 25.

$IP_b=7.90$ eV[172] respectively. The photon order was found using the equation:

$$m = (IP_a - IP_b)/(\nu_a - \nu_b).$$

The solution, m=2.15($\pm$0.2), was found showing that resonant states reside at the sum of two photons. Since four photons of energy are needed to ionize both radicals, the REMPI process occurs through a 2+2 mechanism.

Quantum defects of δ=0.92 and of δ=0.91 and electronic origins at ν_{00}=40,085 cm-1 and at ν_{00}=38,369 cm-1 were computed for allyl and 2-methylallyl radicals respectively. These quantum defects support assignments of these states to the 3s $^2A_1$ Rydberg states. Recently Ha et al.[173] reported results of an ab initio configuration interaction calculation of the excited states of allyl radical. In good agreement with these experiments they calculated that the 3s Rydberg state lies at 42,995 cm^{-1}.

This REMPI study was the first to identify the 3s $^2A_1$ Rydberg state. Previous one photon absorption states did not detect these bands. The reason for this fact is simple. Under the C_{2v} symmetry group the 3s $^2A_1$ state is accessible from the vibrationless $\tilde{X}$ $^2A_2$ ground state only through multiphoton transitions. One photon electric dipole operators do not combine with the A_1 and A_2 states.

Hudgens and Dulcey assigned the regular vibrational progression observed in the spectra to the (a_1) C-C-C in-plane bending mode in the 3s $^2A_1$ Rydberg state. This bending frequency is 390, 342, and 432 cm^{-1} in allyl, allyl-d_6, and 2-methylallyl radicals respectively.

A recent paper by Sappey and Weisshaar[25] has described REMPI experiments which examined vibrationally cold allyl radicals produced by 193 nm photolysis of C_3H_5Cl in a supersonic beam. They found that the REMPI spectrum is very sensitive to the pulse energy and focusing conditions of the ionizing laser. Figure 39 compares the allyl spectrum observed at 6.8 mJ/pulse of energy focused with a 10 cm lens and the allyl spectrum observed at 2.5 mJ and a 15 cm lens. The lower intensity conditions display sharp vibrational bands. The spectrum observed under more intense conditions is broadened on the blue side of each vibrational band. Sappey and Weisshaar have suggested that laser induced interactions of the third laser photon with high lying Rydberg levels may induce this spectral broadening.

Table 25 lists the spectral features observed by Weisshaar in molecular beams. The higher resolution of this data improves the measurements of the allyl radical vibrational band positions and origin (ν_{00}=40,056.8 cm^{-1}). Evidence for three active vibrational modes in the ground state and in the excited state are presented. The (a_1) C-C-C bending frequency was found to be 393 cm^{-1} in the 3s $^2A_1$ state and 426 cm^{-1} in the $\tilde{X}$ $^2A_2$ state. Other firm assignments are pending until the completion of studies with isotopically substituted analogues. The $\tilde{X}$ $^2A_2$ state (a_1) C-C-C bending frequency found by Sappey and Weisshaar is in good agreement with the frequency, 476 cm^{-1}, calculated by Takada and Dupuis[168] during an *ab initio* molecular orbital calculation. Such *ab initio* frequencies are typically 10-15% higher than experimentally observed.[168]

2. Substituted Allyl Radicals

Hudgens et al. have studied the REMPI spectra of the products of the reactions of chlorine and fluorine with cis-2-butene and trans-2-butene between laser wavelengths of 465-485 nm. These reactions conserve the orientation of the methyl group to produce the cis-2-butene-1-yl radical and cis-2-butene-1-yl radical. Laser ionization of these radicals generates only the molecular ion, m/z 55. The REMPI spectrum of the cis-2-butene-1-yl radical exhibits a broad band centered at 472.5 nm and a less intense broad band centered at ~464 nm. The spectrum of trans-2-butene-1-yl shows poorly resolved broad bands centered at 468, 471.5, 474.5, 477.2, 481, and 483.7 nm.

The photon order of the resonant state has not been determined. The ionization potential of these radicals is 7.49 eV.[172] Since three laser photons can ionize the radicals, either a 1+2 or a 2+1 REMPI mechanism can account for the spectra.

3. Benzyl Radical

At present the benzyl radical is the only aromatic free radical for which any REMPI spectrum is known.[27] The benzyl radical appears in the degradation of aromatic fuels which possess an aliphatic carbon side chain,[174] e.g. ethylbenzene. Previous one photon UV studies have reported three electronic absorption bands and deduced the existence of a fourth electronic state in the benzyl radical.[175-180] Benzyl radical can be detected by LIF spectroscopy.[181-182] No electronic states residing at energies higher than 43,300 cm^{-1} are known. Benzyl radical is of C_{2v} symmetry and possesses a $\tilde{X}\ ^2B_2$ ground state.

During the REMPI studies reported by Hoffbauer and Hudgens,[27] benzyl radicals were produced by the hydrogen abstraction reactions, F + toluene and Cl + toluene (Figure 41). The REMPI spectrum appeared only at m/z 91. Since the benzyl radical is the only exothermic reaction channel common to both halogen abstraction reactions, the m/z 91 carrier of the spectra in Figure 41 was assigned to benzyl radicals. Consistent with this assignment of the spectrum to benzyl radicals, the reaction of Cl + toluene-d_8 generated a spectrum carried only by m/z 98.

The benzyl spectrum displays bands at 502.5, 504.8, 506.6 and ~510-515 nm. The benzyl-d_7 spectrum displays bands at 496.2, 502.5, 505.0, and 508-515 nm. When this work was reported, Hoffbauer and Hudgens proposed three

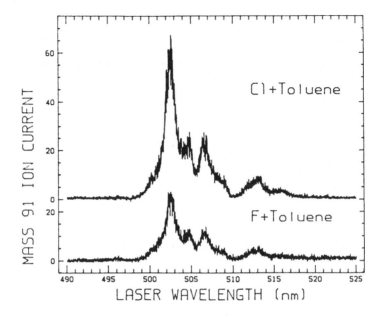

<u>Figure 41.</u> The m/z 91 (benzyl cation) signal observed from the flow reactor effluent between 490-525 nm for the reactions (top) Cl + toluene and (bottom) F + toluene. From Ref. 27.

different excitation mechanisms which could have produced the REMPI

spectrum. These mechanisms were:

1) A 2+1 mechanism. Two photon absorption prepared a low-lying Rydberg state. One more photon was absorbed to generate a ground state cation (IP= 7.2 eV[171]). A total of three photons ionize the radical.

2) A 2+2 mechanism. Two photon absorption prepared the
3 $^2B_2 \leftrightarrow \tilde{X}$ $^2B_2$ valence transition (previously observed between
245-260 nm by one photon spectroscopy). The 3 $^2B_2$ state
corresponds to three singly occupied π-orbitals. To ionize, the
radical must absorb two more photons (4 photons total). This
process generates electronically excited cations
(I.P.=~9.5-9.7 eV[171]).

3) A 3+1 mechanism. Three photon absorption promoted an electron
from the highest <u>fully occupied</u> π-orbital into a Rydberg state
which possesses an ion core with the same electron configuration
as the electronically excited cation. Absorption of one more
photon generated the electronically excited cation (IP=~9.5-9.7
eV). A total absorption of four photons ionized the radical.

Hoffbauer and Hudgens favored mechanism 2), but were unable to offer a

conclusive assignment because the cited one photon UV absorption spectrum of

the 3 $^2B_2$ state was observed from vibrationally hot benzyl radicals.[177] The

REMPI spectrum viewed relatively cold benzyl radicals. Since vibrational

temperature influences band contours, comparisons between the two spectra,

although suggestive, were inappropriate for rendering an assignment.

A suitable UV spectrum of the 3 $^2B_2$ state observed in vibrationally cold

benzyl radicals was recently reported by Ikeda et al.[183] When the UV spectrum

between 245-260 nm and the (presumed two photon) REMPI spectrum between 490-

520 nm are compared, their principal features appear at the same energy with

very similar band envelopes. This evidence strongly favors assignment of the

REMPI spectrum to two photon absorption into the 3 $^2B_2$ state.

Since the 3 $^2B_2$ state corresponds to three singly occupied π-orbitals,

ionization will produce an electronically excited benzyl cation. The lowest

electronic state of benzyl cation is estimated to lie 9.5-9.7 eV above the

benzyl radical ground state. Thus, to ionize, the 3 $^2B_2$ state benzyl radical

must absorb two more photons (4 photons total).

From this brief study of benzyl radical we can conclude that aromatic ring systems will probably be more difficult to reach conclusive state assignments because a greater number of plausible REMPI mechanisms appear within the same spectral region. The fact that several REMPI excitation mechanisms must be considered --including those which involve the preparation of Rydberg states with cation cores in electronically excited states--will be typical of studies involving aromatic ring systems.

4. Cyclohexanyl radicals.

The hydrogen abstraction reaction of fluorine and chlorine atoms with cyclohexane produces an m/z 83 ion signal which Hudgens et al.[28] have attributed to the cyclohexanyl radical. Figure 6 above shows the REMPI mass spectrum of this radical at 500.6 nm. The optical spectrum is a featureless continuous band which follows the laser dye curve between 460-520 nm. The REMPI signal increases and decreases are a function of halogen atom and cyclohexane concentration. The heat of formation of cyclohexanyl radical is $\Delta H_f^o = 175$ kcal/mole which suggests and ionization potential of 6.8 eV.[184] Thus, ionization could occur through a three photon ionization mechanism.

5. Ethyl Radical

Sappey and Weisshaar[25] have observed the REMPI signals from ethyl radicals produced in a molecular beam. The radicals were produced by 193 nm photolysis of $C_2H_5NO_2$ in an extension tube attached to a pulsed molecular beam valve. The photolysis products were cooled by supersonic expansion in a 2.5 atm back pressure of argon from the valve.

Wendt and Hunziker[185] have recently reported a relatively featureless one photon absorption spectrum of C_2H_5 radicals around 205 nm which were assigned to $3p \leftarrow \tilde{X}$ $^2A_1$ transitions. Based upon this observation, Sappey and Weisshaar[25] irradiated the C_2H_5 radicals produced in the supersonic expansion with 398.5-409.5 nm, 2.5 mJ/pulse laser light focused with a 10 cm lens. These conditions should ionize C_2H_5 by a 2+1 REMPI mechanism through the excited state reported by Wendt and Hunziker. They observed very small ionization signals ($^\sim$ 1 ion/pulse) at the parent ion mass, m/z 29. The signal was dependent on the precursor and the photolysis laser, but no spectral features were resolved.

Sappey and Weisshaar suggested that the small REMPI signal strength was due to rapid destruction of the nascent C_2H_5 radicals in the source. This explanation is reasonable because the activation energy for hydrogen atom loss from ethyl radical is only 38 kcal/mole.[186] Thus, because laser photolysis generally produces hot radicals which may rapidly decompose before supersonic cooling can effectively quench rapid dissociation, a low yield of C_2H_5 is not surprising.

6. Methoxy Radical

The methoxy radical is probably the most studied of all polyatomic radicals.[187] It appears in reaction sequences of combustion, atmospheric, and interstellar chemistry. Methoxy radical is also an intermediate in the CO/H_2 chemistry on surfaces. The methoxy radical $\tilde{A}$ $^2A_1$-$\tilde{X}$ ^2E band system has been well-studied by absorption and fluorescence spectroscopies.[187-192] and the origin lies near 309.6 nm ($^\sim$32,300 cm^{-1})[187] LIF spectroscopy is often used for kinetic studies which involve methoxy radicals.[190-192]

While conducting the reaction of F + methanol in a flow reactor, Long et al.[23] observed a m/z 31 REMPI spectrum between 315-330 nm (Figure 42a). When they conducted the reaction, F + CD_3OD, the spectrum shifted to m/z 34 which confirmed that the spectral carriers possessed three hydrogen atoms. Since the F + methanol reaction produces both methoxy and hydroxymethyl radicals, a REMPI spectrum of the products from the reaction of fluorine with the partially substituted reagent, CH_3OD, was conducted to resolve between these possible spectral carriers. As shown in Figure 42b, the reaction of F + CH_3OD produced m/z 31 REMPI spectrum between 315-327 nm which was identical to the m/z 31 spectrum observed from the reaction of F + CH_3OH. Thus, the data proved that the spectral carrier is methoxy radical formed by abstraction of the hydroxyl hydrogen.

The spectra consisted of dense packets of rotational lines which caused them to appear noisy and congested. The principal bands appeared at 316.9, 320.6, 324.1, and 326.2 nm in CH_3O and at 314.0, 314.8, 316.0, 316.8, 319.2, 322.2, 324.1, and 325.8 nm in CD_3O. Long et al. reported that the sensitivity for methoxy radicals was within an order of magnitude of that for hydroxymethyl radicals.

The fact that only the molecular ion (m/z 31 & 34) was observed in the REMPI mass spectra is significant. Calculations[193-194] and experiments[195] show that the ground state $\tilde{X}$ $^1A_1$ methoxy cation rapidly dissociates into fragments. This fact led Long et al.[23] to conclude that the REMPI excitation process generates a long-lived, excited state of the methoxy cation.

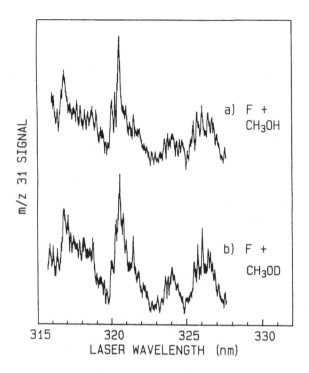

<u>Figure 42.</u> The REMPI spectrum of methoxy radical between 315-328 nm from the reactions: a) F + CH$_3$OH and b) F + CH$_3$OD. From Ref. 23.

Reprinted with permission from J. Phys. Chem. 90, 4901 (1986) Copyright (1986) American Chemical Society

Calculations have shown that the ã ^{3}A$_1$ methoxy cation is stable.[194] Charge reversal experiments by Burgers and Holmes[196] recently reported ΔH_f^o(CH$_3$O$^+$ ã ^{3}A$_1$)=1034(+/-20) kJ/mole which leads to a derived ionization potential of 10.5 eV for CH_3O ($\tilde{X}\ ^2E$).[197] Based upon this result, Long et al.

proposed two possible mechanisms which could account for the REMPI spectrum:

1) A 2+1 REMPI mechanism. Simultaneous two photon absorption excites the 3s $^2A_1$ Rydberg state. The Rydberg radicals may ionize and become $\tilde{a}\ ^3A_1$ cations after they absorb one additional laser photon. To date, no Rydberg states of methoxy radical have been reported.

2) A 1+2 or 1+3 REMPI mechanism. One photon absorption prepares the $\tilde{A}\ ^2A_1$ valence state of the methoxy radical. The $\tilde{A}\ ^2A_1$ methoxy radicals must absorb at least two additional laser photons to produce electronically excited cations, e.g. $\tilde{b}\ ^3E$ state cations. If these excited cation states lie greater than 11.2 eV relative to the methoxy radical, then ionization of ground state methoxy radicals require absorption of a total of four laser photons.

Subsequent data[198] recorded between 275-390 nm shows that the band

positions within the REMPI spectrum do not duplicate those reported in

absorption or fluorescence spectra of the methoxy radical $\tilde{A}\ ^2A_1 \leftarrow \tilde{X}\ ^2E$ band

system. Although no firm assignment is available from this work, the REMPI

mechanism does not appear to involve a valence state transition.

ACKNOWLEDGMENTS

The author thanks Drs. Marilyn E. Jacox and Russell D. Johnson of the

National Bureau of Standards (Gaithersburg, MD) for discussions, suggestions,

and reading of the manuscript during the preparation of this review. The

author also thanks Prof. T. A. Cool (Cornell University) and Prof. J. C.

Weisshaar (University of Wisconsin) for preprints.

V. REFERENCES

1. P. M. Johnson, M. R. Berman, and D. Zakheim, J. Chem. Phys. 62, 2500 (1975).

2. G. Petty, C. Tai, and F. W. Dalby, Phys. Rev. Lett. 34, 1207 (1975).

3. G. C. Nieman and S. D. Colson, J. Chem. Phys. 68, 5656 (1978).

4. J. H. Glownia, S. J. Riley, S. D. Colson, and G. C. Nieman, J. Chem. Phys. 73, 4296 (1980).

5. T. G. DiGiuseppe, J. W. Hudgens, and M. C. Lin, Chem. Phys. Lett. 82, 267 (1981).

6. T. G. DiGiuseppe, J. W. Hudgens, and M. C. Lin, J. Phys. Chem. 86, 36 (1982).

7. M. T. Duignan, J. W. Hudgens, and J. R. Wyatt, J. Phys. Chem. 86, 4156 (1982).

8. J. Danon, H. Zacharias, H. Rottke, and K. H. Welge, J. Chem. Phys. 76, 2399 (1982).

9. T. G. DiGiuseppe, J. W. Hudgens, and M. C. Lin, J. Chem. Phys. 76, 3338 (1982).

10. J. W. Hudgens, T. G. DiGiuseppe, and M. C. Lin, J. Chem. Phys. 79, 571 (1983).

11. M. T. Duignan and J. W. Hudgens, J. Chem. Phys. 82, 4426 (1985).

12. S. Sharpe and P. M. Johnson, Chem. Phys. Lett. 107, 35 (1984).

13. P. Chen, W. A. Chupka, and S. D. Colson, Chem. Phys. Lett. 121, 405 (1985).

14. R. D. Johnson and J. W. Hudgens, to appear in J. Phys. Chem 91,   (1987).

15. M. N. R. Ashfold, R. N. Dixon, and R. J. Strickland, Chem. Phys. Lett. 111, 226 (1985).

16. K. C. Smyth and W. G. Mallard, J. Chem. Phys. 77, 1779 (1982).

17. C. S. Dulcey and J. W. Hudgens, Chem. Phys. Lett. 118, 444 (1985).

18. P. J. H. Tjossem and T. A. Cool, 20th Symposium (Internat.) on Combust. (The Combust. Inst., Pittsburg, 1984), p. 1321.

19. P. J. H. Tjossem, P. M. Goodwin, and T. A. Cool, J. Chem. Phys. 84, 5334 (1986).

20. J. A. Dagata, D. W. Squire, C. S. Dulcey, D. S. Y. Hsu, and M. C. Lin, Chem. Phys. Lett. 134, 151 (1987).

21. J. W. Hudgens, C. S. Dulcey, G. R. Long, and D. J. Bogan, to appear in J. Chem. Phys. 87, (1987).

22. C. S. Dulcey and J. W. Hudgens, J. Chem. Phys. 84, 5262 (1986).

23. G.R. Long, R.D. Johnson, and J.W. Hudgens, J. Phys.Chem. 90, 4901 (1986).

24. G. R. Long and J. W. Hudgens, to appear in J. Phys. Chem. 91, (1987).

25. A. D. Sappey and J. C. Weisshaar, J. Phys. Chem. 91, 3731 (1987).

26. J. W. Hudgens and C. S. Dulcey, J. Phys. Chem. 89, 1505 (1985).

27. M. A. Hoffbauer and J. W. Hudgens, J. Phys. Chem. 89, 5152 (1985).

28. J. W. Hudgens, G. R. Long, and R. D. Johnson, in press.

29. J. C. Miller and W. C. Cheung, J. Phys. Chem. 89, 1643 (1985).

30. G. Placzek and Teller, Z. Phys. 81, 209 (1933).

31. F. Metz, W. E. Howard, L. Wunch, H. J. Neusser, and E. W. Schlag, Proc. R. Soc. A. 363, 381 (1978).

32. W. M. McClain, Acc. Chem. Res. 7, 129 (1974).

33. W. M. McClain and R. A. Harris, Excited States, ed. E. C. Lim (Academic, New York, 1977), Vol. III.

34. K. Chen and E. S. Yeung, J. Chem. Phys. 69, 43 (1978). See Ref. 9 for correction of a typographical error in eq. 15.

35. J. B. Halpern, H. Zacharias, and R. Wallenstein, J. Mol. Spectrosc. 79, 1 (1980). Consult JBH for errata.

36. R. G. Bray and R. M. Hochstrasser, Mol. Phys. 31, 1199 (1976). For erratum: A. C. Kummel, G. O. Sitz, and R. N. Zare, J. Chem. Phys. 85, 6874 (1986).

37. R. N. Dixon, J. M. Bayley, and M.N.R. Ashfold, Chem. Phys. 84, 21 (1984).

38. G. C. Nieman, J. Chem,. Phys. 75, 584 (1981).

39. Y. N. Chiu, J. Chem. Phys. 86, 1686 (1987).

40. M. N. R. Ashfold, R. N. Dixon, J. D. Prince, and B. Tutcher, Mol. Phys. 56, 1185 (1985).

41. D. L. Andrews and W. A. Ghoul, J. Chem. Phys. 75, 530 (1981).

42. G. Herzberg, *Molecular Spectra and Molecular Structure* (Van Nostrand-Reinhold, New York, 1966), Vol. III.

43. A. B. F. Duncan, *Rydberg Series in Atoms and Molecules* (Academic Press, New York, 1971).

44. M. B. Robin, *Higher Excited States of Polyatomic Molecules* (Academic, New York, 1975).

45. S. T. Manson, Phys. Rev. $\underline{182}$, 97 (1969).

46. U. Fano, C. E. Theodosiou, and J. L. Dehmer, Rev. Mod. Phys. $\underline{48}$, 49 (1976).

47. G. K. Anderson, and S. H. Bauer, J. Phys. Chem. $\underline{81}$, 1146 (1977).

48. C. S. Dulcey and J. W. Hudgens, J. Phys. Chem. $\underline{87}$, 2296 (1983).

49. R. McDiarmid, J. Chem. Phys. $\underline{55}$, 2426 (1971).

50. D. A. Gobeli, J. J. Yang, and M. A. El-Sayed, Chem. Rev. $\underline{85}$, 529 (1985).

51. W. P. White, Jr., Dissertation (Ohio State Univ., 1971).

52. K. P. Huber and G. Herzberg, *Constants of Diatomic Molecules*, (Van Nostrand-Reinhold, New York, 1976).

53. W. P. White, R. M. Pitzer, C. W. Mathews, and T. H. Dunning, J. Mol. Spectrosc. $\underline{75}$, 4287 (1982).

54. J. M. Dyke, A. E. Lewis, and A. Morris, J. Chem. Phys. $\underline{80}$, 1382 (1984).

55. A. Carrington and B. J. Howard, Mol. Phys. $\underline{18}$, 225 (1970).

56. J. W. Hepburn, D. J. Trevor, J. E. Pollard, D. A. Shirley, and Y. T. Lee, J. Phys. Chem. $\underline{76}$, 4287 (1982).

57. J. M. Brown, J. E. Schubert, R. J. Saykally, and K. M. Evenson, J. Mol. Spectrosc. $\underline{120}$, 421 (1986).

58. T. L. Porter, D. E. Mann, and N. Acquista, J. Mol. Spectrosc. $\underline{16}$, 228 (1965).

59. C. E. Moore and H. P. Broida, J. Res. Natl. Bur. Stand. Sect. A $\underline{63}$, 19 (1959).

60. A. E. Douglas and G. A Elliot, Can. J. Phys. $\underline{43}$, 496 (1965).

61. I. Botterud, A. Loftus, and L. Veseth, Phys. Scripta $\underline{8}$, 218 (1973).

62. S. V. Filseth, H. Zacharias, J. Danon, R. Wallenstein, K. H. Welge, Chem. Phys. Lett. $\underline{58}$, 140 (1978).

63. G. Herzberg and J. W. C. Johns, Astrophys. J. 158,.399 (1969).

64. J. B. Pallix, P. Chen, W. A. Chupka, and S. D. Colson, J. Chem. Phys. 84, 5208 (1985).

65. P. Chen, J. B. Pallix, W. A. Chupka, and S. D. Colson, J. Chem. Phys. 86, 516 (1986).

66. G. Pannetier and A. G. Gaydon, Nature 161, 242 (1948).

67. W. M. Vaidya, Proc. Indian Acad. Sci. A 6, 122 (1937).

68. W. M. Vaidya, Proc. Indian Acad. Sci. A 7, 321 (1938).

69. E. H. Coleman and A. G. Gaydon, Discuss. Faraday Soc. 2, 166 (1947).

70. R. A. Durie and D. A. Ramsay, Can. J. Phys. 36, 35 (1958).

71. N. Basco and R. D. Morse, J. Mol. Spectrosc. 45, 35 (1973).

72. A. G. Maki, F. J. Lovas, and W. B. Olson, J. Mol. Spectrosc. 92, 410 (1982).

73. J. A. Coxon, Can. J. Phys. 57, 1538 (1979).

74. D. K. Bulgin, J. M. Dyke, N. Johnathan, and A. Morris, Mol. Phys. 32, 1487 (1976).

75. D. K. Bulgin, J. M. Dyke, N. Johnathan, and A. Morris, J. Chem. Soc. Faraday Trans. 2 75, 456 (1978).

76. A. R. W. McKellar, J. Mol. Spectrosc. 86, 43 (1981).

77. J. E. Butler, K. Kawaguchi, and E. Hirota, J. Mol. Spectrosc. 104, 372 (1984).

78. S. J. Dunlavey, J. M. Dyke, and A. Morris, Chem. Phys. Lett. 53, 382 (1978).

79. J. A. Coxon and D. A. Ramsay, Can. J. Phys. 54, 1034 (1976).

80. J. A. Coxon, W. E. Jones, and E. G. Skolnik, Can. J. Phys. 54, 1043 (1976).

81. M. Barnett, E. A. Cohen, and D. A. Ramsay, Can. J. Phys. 59, 1908 (1981).

82. J. Berkowitz,, L. Curtiss, S. Gibson, J. Greene, G. Hillhouse, and J. Pople, J. Chem. Phys. 84, 375 (1986).

83. J. Rostas, D. Cossart, and J. B. Bastien, Can. J. Phys. 52, 1274 (1974).

84. A. T. Droege and P. C. Engelking, J. Chem. Phys. 80, 5926 (1984).

85. J. W. Hastie and D. W. Bonnell, "Molecular Chemistry of Inhibited Combustion Systems," Natl. Bur. Stand. IR 80-2169, October (1980).

86. K. N. Wong, W. R. Anderson, A. J. Kotlar, M. A. DeWilde, and L. J. Decker, J. Chem. Phys. 84, 81 (1986).

87. C. Couet, N. T. Anh, B. Coquart, and H. Guenebaut, J. Chim. Phys. 65, 217 (1968).

88. R. D. Verma and S. R. Singhal, Can. J. Phys. 53, 411 (1975).

89. T. V. R. Rao, R. R. Reddy, and P. S. Rao, Physica C 106, 445 (1981).

90. S. J. Davis and S. G. Hadley, Phys. Rev. A14, 1146 (1976).

91. R. F. Barrow, D. Butler, J. W. C. Johns, and J. L. Powell, Proc. Roy. Soc. 73, 319 (1959).

92. Y. Houbrechts, I. Dubois, and H. Bredohl, J. Phys. B15, 603 (1982).

93. O. Appelblad, R. F. Barrow, and R. D. Verma, J. Phys. B1, 274 (1968).

94. R. W. Martin and A. J. Merer, Can. J. Phys. 51, 634 (1973).

95. J. W. C. Johns and R. F. Barrow, Proc. Roy. Soc. 71, 476 (1958).

96. V. M. Donnelley, W. M. Pitts, and J. R. McDonald, Chem. Phys. 49, 289 (1980).

97. M. Pitts, V. M. Donnelley, A. P. Baronavski, and J. R. McDonald, Chem. Phys. 51, 451 (1981).

98. O. Horie, W. Bauer, R. Meuser, V. H. Schmidt, and K. H. Becker, Chem. Phys. Lett. 100, 251 (1983).

99. K. Dressler and D. A. Ramsay, Philos. Trans. R. Soc. London Ser. A. 251, 553 (1959).

100. J. H. Clark, C. B. Moore, and J. P. Reilly, Int. J. Chem. Kinet. 10, 427 (1978).

101. J. P. Reilly, J. H. Clark, C. B. Moore, and G. C. Pimentel, J. Chem. Phys. 69, 4381 (1978).

102. R. J. Gill and G. H. Atkinson, Chem. Phys. Lett. 64, 426 (1979).

103. B. M. Stone, M. Noble, and E. K. C. Lee, Chem. Phys. Lett. 118, 83 (1985).

104. J. M. Dyke, N. B. H. Johnathan, A. Morris, and M. J. Winter, Mol. Phys. 39, 629 (1973).

105. T. A. Cool, private communication.

106. K. K. Murray, T. M. Miller, D. G. Leopold, and W. C. Lineberger, J. Chem. Phys. $\underline{84}$, 2520 (1986).

107. C. S. Gudeman, M. H. Begemann, J. Pfaff, and R. J. Sakally, Phys. Rev. Lett. $\underline{50}$, 727 (1983); (b) T. Amano, J. Chem. Phys. $\underline{79}$, 3595 (1983).

108. J. W. C. Johns, S. H. Priddle, and D. A. Ramsay, Discuss. Faraday. Soc. $\underline{35}$, 90 (1963).

109. B. M. Landsberg, A. J. Merer, and T. Oka, J. Mol. Spectrosc. $\underline{67}$, 459 (1977).

110. J. W. C. Johns, A. R. W. McKellar, and M. Riggin, J. Chem. Phys. $\underline{67}$, 2427 (1977).

111. P. B. Davies and W. J. Rothwell, J. Chem. Phys. $\underline{81}$, 5239 (1984).

112. R. S. Lowe and A. R. W. McKellar, J. Chem. Phys. $\underline{74}$, 2686 (1981).

113. P. B. Davies, P. A. Hamilton, and W. J. Rothwell, J. Chem. Phys. $\underline{81}$, 1598 (1984).

114. K. Kawaguchi, A. R. W. McKellar, and E. Hirota, J. Chem. Phys. $\underline{84}$, 1146 (1986).

115. J. M. Dyke, J. Chem. Soc. Faraday Trans 2 $\underline{83}$, 69 (1987).

116. S. C. Foster and A. R. W. McKellar, J. Chem. Phys. $\underline{81}$, 3424 (1984).

117. J. H. Gole, R. H. Hague, J. L. Margrave, J. W. Hastie, J. Mol. Spectrosc. $\underline{43}$, 441 (1972).

118. A. C. Stanton, A. Freedman, J. Wormboudt, and P. P. Gaspar, Chem. Phys, Lett. $\underline{122}$, 190 (1985).

119. Y. Matsumi, S. Toyoda, T. Hayashi, M. Miyamura, H. Yoshikawa, and S. Komiya, J. Appl. Phys. $\underline{60}$, 4102 (1986).

120. T. P. Fehlner and D. W. Turner, Inorg. Chem. $\underline{13}$, 754 (1974).

121. N. P. C. Westwood, Chem. Phys. Lett. $\underline{25}$, 558 (1974).

122. C. Yamada, E. Hirota, and K. Kawaguchim J. Chem. Phys. $\underline{75}$, 5256 (1981).

123. M. Karplus, J. Chem. Phys. $\underline{30}$, 15 (1959).

124. R. McDiarmid, Theor. Chim. Acta (Berlin) $\underline{20}$, 382 (1971).

125. B. Lengsfeld III, P. E. M. Siegbahn, and B. Liu, J. Chem. Phys. $\underline{81}$, 710 (1984).

126. G. Herzberg and J. Shoosmith, Can. J. Phys. $\underline{34}$, 523 (1956).

127. G. Herzberg, Proc. Roy. Soc. London Ser. A $\underline{262}$, 291 (1961).

128. (a) T. Koenig, T. Balle, and W. Snell, J. Am. Chem. Soc. $\underline{97}$, 662 (1975);
     (b) T. Koenig, T. Balle, and J. C. Chang, Spectrosc. Lett. $\underline{9}$, 755
     (1976).

129. J. Dyke, N. Jonathan, E. Lee, and A. Morris, J. Chem. Soc., Faraday
     Trans. II, $\underline{72}$, 1385 (1976).

130. P. L. Holt, K. E. McCurdy, R. B. Weisman, J. S. Adams, and P. S. Engel,
     J. Chem. Phys. $\underline{81}$, 3349 (1984).

131. D. J. DeFrees and A. D. McLean, J. Chem. Phys. $\underline{82}$, 333 (1985).

132. L. Y. Tan, A. M. Winer, and G. C. Pimentel, J. Chem. Phys. $\underline{57}$, 4028
     (1972).

133. A. B. Callear and M. P. Metcalfe, Chem. Phys. $\underline{14}$, 275 (1976).

134. (a) D. E. Milligan and M. E. Jacox, J. Chem. Phys. $\underline{47}$, 5146 (1967); (b)
     M. E. Jacox, J. Mol. Spectrosc. $\underline{66}$, 272 (1977).

135. K. C. Smyth and P. H. Taylor, Chem. Phys. Lett. $\underline{122}$, 518 (1985).

136. P. Chen, S. D. Colson, W. A. Chupka, and J. A. Berson, J. Phys. Chem.
     $\underline{90}$, 2319 (1986).

137. L. Andrews, J. M. Dyke, N. Jonathan, N. Keddar, A. Morris, and A. Ridha,
     J. Chem. Phys. $\underline{88}$, 2364 (1984).

138. (a) E. B. Wilson, Jr., J. C. Decius, and P. C. Cross, Molecular
     Vibrations, (McGraw-Hill, New York, 1955). (b) J. H. Schachtschneider,
     "Vibrational Analysis of Polyatomic Molecules", Tech. Report Nos. 231-64
     and 57-65 (Shell Development Co., Emeryville, 1964).

139. C. Yamada and E. Hirota, J. Mol. Spectrosc. $\underline{116}$, 101 (1986).

140. (a) M. E. Jacox and D. E. Milligan, J. Chem. Phys. $\underline{50}$, 3253 (1969).
     (b) J. I. Raymond and L. Andrews, J. Chem. Phys. $\underline{75}$, 3225 (1971).

141. M. E. Jacox, Chem. Phys. $\underline{59}$, 199 (1981).

142. S. W. Benson and M. Weissman, Int. J. Chem. Kinet. $\underline{14}$, 1287 (1982).

143. M. E. Jacox and D. E. Milligan, J. Chem. Phys. $\underline{54}$, 3935 (1971).

144. S. P. Mishra, G. W. Neilson, and M. C. R. Symons, J. Chem. Soc. Faraday
     Trans. 2, $\underline{69}$, 1425 (1973).

145. L. Andrews, J. M. Dyke, N. Jonathan, N. Keddar, and A. Morris, J. Chem. Phys. 79, (1983), 4650.

146. L. M. Molino, J. M. Poblet, and E. Canadell, J. Chem. Soc. Perkin Trans. II, (1982) 1217.

147. (a) G. LeBras, N. I. Butkovskaya, I. I. Morozov, and V. L. Talrose, Chem. Phys. 50, 63 (1980). (b) M. A. Nazar and J. C. Polanyi, Chem. Phys. 55, 299 (1981).

148. D. S. Bomse, S. Dougal, and R. L. Woodin, J. Phys. Chem. 90, 2640 (1986).

149. M. E. Jacox, Chem. Phys. 59, 213 (1981).

150. D. Solgadi and J. P. Flament, Chem. Phys. 98, 387 (1985).

151. J. M. Dyke, A. R. Ellis, N. Jonathan, N. Kedder, and A. Morris, Chem. Phys. Lett. 111, 207 (1984).

152. "Multiphoton Ionization of Radicals: Pyrolysis and Infrared Multiphoton Generated CF$_3$ and CH$_3$", M. T. Duignan, T. G. DiGiuseppe, J. W. Hudgens and J. R. Wyatt, Lasers as Reactants and Probes in Chemistry, W. M. Jackson and A. B. Harvey, eds., Howard University Press, Washington, D. C. (1985), p. 113.

153. N. Basco and F. G. M. Hathorn, Chem. Phys. Lett. 8, 291 (1971).

154. C. Lifshitz and W. A. Chupka, J. Chem. Phys. 47, 2704 (1965).

155. R. W. Fessenden and R. H. Schuler, J. Chem. Phys. 43, 2704 (1965).

156. N. Washida, M. Suto, S. Nagase, U. Nagashima, and K. Morokuma, J. Chem. Phys. 78, 1025 (1983).

157. G. A. Carlson and G. C. Pimentel, J. Chem. Phys. 44, 4058 (1966).

158. M. Suto and N. Washida, J. Chem. Phys. 78, 1012 (1983).

159. W. Tsang, Int. J. Chem. Kin. 10, 1119 (1978).

160. A. Tiesler, H. Zatt, H. Heusinger, Radiat. Phys. Chem 17, (1984).

161. R. E. Linder, D. L. Winters, and A. C. Ling, Can. J. Chem. 54, 1405 (1976).

162. D. J. Driscoll and J. H. Lunsford, J. Phys. Chem. 103, 301 (1983).

163. J. C. Schultz and J. L. Beauchamp, J. Phys. Chem. 87, 3587 (1983).

164. R. W. Fessenden and R. H. Schuler, J. Phys. Chem. 39, 2147 (1968).

165. G. Levin and W. A. Goddard III, Theor. Chim. Acta, $\underline{37}$, 253 (1975).

166. G. Levin and W. A. Goddard III, J. Am. Chem. Soc. $\underline{97}$, 1649 (1975).

167. A. F. Voter and W. A. Goddard III, Chem. Phys. $\underline{57}$, 253 (1981).

168. T. Takada and M. Dupuis, J. Am. Chem. Soc. $\underline{105}$, 1713 (1983).

169. C. L. Currie and D. A. Ramsay, J. Chem. Phys. $\underline{45}$, 488 (1966).

170. A. B. Callear and H. K. Lee, Trans. Faraday Soc. $\underline{64}$, 308 (1968).

171. F. A. Houle and J. L. Beauchamp, J. Am. Chem. Soc. $\underline{100}$, 3290 (1978).

172. J. C. Schultz, F. A. Houle, J. L. Beauchamp, J. Am. Chem. Soc. $\underline{106}$, 7336 (1984).

173. T. K. Ha, H. Baumann, and J. F. M. Oth, J. Chem. Phys. $\underline{85}$, 1438 (1986).

174. K. Brezinsky, Prog. Energy Combust. Sci. $\underline{12}$, 1 (1986).

175. C. Cossart-Magos and S. Leach, J. Chem. Phys. $\underline{56}$, 1534 (1972).

176. J. Jordan, D. W. Pratt, and D. E. Wood, J. Am. Chem. Soc. $\underline{96}$, 5588 (1974).

177. F. Bayrakceken and J. E. Nicholas, J. Chem. Soc. B $\underline{691}$ (1970).

178. R. V. Lloyd and D. E. Wood, J. Chem. Phys. $\underline{60}$, 2684 (1974).

179. G. Porter and F. Wright, Trans. Faraday Soc. $\underline{51}$, 1469 (1955).

180. G. Porter and E. Strachan, Spectrochim Acta. $\underline{12}$, 299 (1958).

181. D. M. Brenner, G. P. Smith, and R. N. Zare, J. Am. Chem. Soc. $\underline{98}$, 6707 (1976).

182. J. H. Miller and L. Andrews, J. Mol. Spectrosc. $\underline{90}$, 20 (1981).

183. N. Ikeda, N. Nakashima, and K. Yoshihara, J. Am. Chem. Soc. $\underline{107}$, 3381 (1985).

184. S. H. Lias, J. E. Bartmess, J. L. Holmes, R. D. Levin, J. F. Liebman, Gas-phase Ion and Neutral Thermochemistry, Nat. Stand. Ref. Data Ser.(U.S. Dept of Commerce, 1987).

185. H. R. Wendt and H. E. Hunziker, J. Chem. Phys. $\underline{82}$, 717 (1985).

186. W. Tsang and R. F. Hampson, J. Phys. Chem. Ref. Data $\underline{15}$, 1087 (1986).

187. S. D. Brossard, P. G. Carrick, E. L. Chappell, S. C. Hulegaard, and P. C. Engelking, J. Chem. Phys. $\underline{84}$, 2459 (1986) and references cited within.

188. G. Inoue, H. Akimoto and M. Okuda, Chem. Phys. Lett. $\underline{63}$, 213 (1979).

189. G. Inoué, H. Akimoto and M. Okuda, J. Chem. Phys. $\underline{72}$, 1769 (1980).

190. N. Sanders, J.E. Butler, L. R. Pasternack and J. R. McDonald, Chem. Phys. $\underline{48}$, 203 (1980).

191. D. J. Bogan, B. E. Brauer, C. W. Hand, M. J. Kaufman, and W. A. Sanders, "Measurement of the Fractional Yield of Methoxy from the Reaction F plus Methanol by Laser Induced Fluorescence and Chemiluminescence", paper E-28, Int. Conf. on Chem. Kinetics, Nat. Bur. of Standards, Gaithersburg, MD, 17-19 June 1985.

192. D. E. Powers, J. B. Hopkins, and R. E. Smalley, J. Phys. Chem. $\underline{85}$, 2711 (1981).

193. M. A. Haney, J. C. Patel, and E. F. Hayes, J. Chem. Phys. $\underline{53}$, 4105 (1970).

194. W. J. Bouma, R. H. Nobes, and L. Radom, Org. Mass Spectrom. $\underline{17}$, 315 (1982).

195. J. D. Dill, C. L. Fischer, and F. W. McLafferty, J. Am. Chem. Soc. $\underline{101}$, 6531 (1979).

196. P. C. Burgers and J. L. Holmes, Org. Mass Spectrom. $\underline{19}$, 452 (1984).

197. For discussion purposes state symmetries are referenced to a C_{3v} molecular symmetry. Evidence indicates that symmetry spoiling Jahn-Teller interactions in methoxy radical are small. See Ref. 187 and D. R. Yarkony, H. F. Schaefer III, and S. Rothenberg, J. Am Chem. Soc. $\underline{96}$, 656 (1974).

198. G. R. Long, R. D. Johnson, and J. W. Hudgens, in press.

Time-Resolved Resonance Raman Spectroscopy: Intense Electromagnetic Field Effects, and Photoelectrochemical Reactions on Semiconductor Crystallite Surfaces

Louis Brus

AT&T Bell Laboratories
Murray Hill, New Jersey 07974

Contents

1. INTRODUCTION

The application of time-resolved resonance Raman scattering to chemical kinetics in homogeneous liquids, and at condensed phase interfaces, has been developing since the first experiments on short-lived free radicals by Wilbrant and coworkers[1] over ten years ago. The motivation for obtaining Raman spectra of transient species is quite clear: Raman spectra are specific fingerprints, and molecular identification is often possible even if multiple species with structureless, overlapping optical spectra are present. A comparison of Stokes and Antistokes intensities allows a determination of vibrational populations on a time-resolved basis. Isotopic Raman spectra provide information on structure, geometry, and chemical bonding. Distortion due to local solvation can sometimes be observed. All of these results help to unravel the mechanisms of complex chemical reactions.

Time-resolved resonance Raman spectroscopy has been thoroughly reviewed by Hub, Schneider, and Dörr in Volume 2 of this series. These authors have outlined Albrecht's Herzberg-Teller theory of the resonance Raman effect, and described in detail various experimental approaches. They also discuss specific results for proteins (e.g., bacteriorhodopsin and hemoglobin), molecular excited states, and various radicals and radical ions (especially those related to stilbene). This article has extensive references to work published before 1986.

In this short chapter I would like to discuss two additional topics. The first is the nature of inelastic (e.g., Raman and fluorescence) resonant scattering at high electromagnetic field intensity. This subject is relevant to time-resolved Raman experiments because intense optical pulses are often utilized on the ns and ps time scales. The second topic is the photoelectrochemical reactions of molecules adsorbed in the double layer on the surface of small colloidal semiconductor crystallites. These systems are prototypes for study of charge transfer across electrode:liquid interfaces. In some case, the crystallites are sufficiently small (less than about 60Å diameter) that quantum size effects occur and the crystallites have an apparent band gap larger than the bulk material.

2. Resonant Scattering in Low and High Fields

The theory of high field, resonant inelastic photon scattering by molecules in condensed phases has been developing over the past decade.[2-8] In order to correctly treat the time evolution of quantum mechanical coherences, a density matrix formalism, rather than a rate equation approach, is required. Instead of presenting density matrix formalism, I simply discuss the phenomena involved so as to create a framework for thinking about resonance Raman and fluorescence in the presence of intense laser fields. These effects in resonance Raman scattering have been experimentally demonstrated just recently in the ground state of β-carotene.

There is a certain hierarchy of energy interactions for dissolved molecules. As in the gas phase, the most important Hamiltonian term is the internal electronic and vibrational motion. This produces, in zero order, a series of electronic states, each with its own vibrational normal modes. Figure 1 shows schematically two S_0 vibronic levels and a resonant S_1 vibronic level. In liquid solution, a molecule additionally undergoes rapid perturbations due to collisions with neighbors, causing time dependent

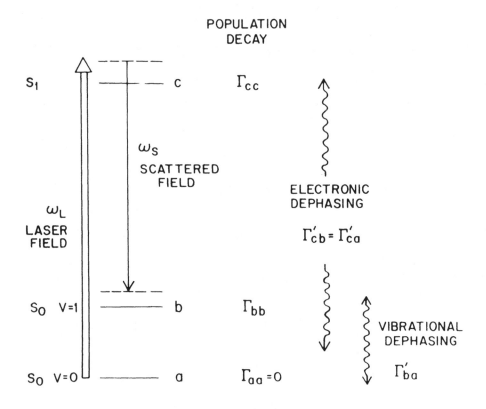

Fig. 1   Schematic illustration of a three level resonance scattering model, incorporating both T_2 and T_1 processes (from reference 9).

vibronic energy fluctuations. Such collisions do not change the populations in the various states. These collisional, time dependent energy fluxuations are termed dephasing (T_2 processes); they contribute to the observed linewidth $\Gamma_{\alpha\beta}$ of any molecular transition between two internal levels α and β.

$$\Gamma_{\alpha\beta} = \frac{1}{2}\left[\Gamma_{\alpha\alpha}+\Gamma_{\beta\beta}\right] + \Gamma'_{\alpha\beta} \tag{1}$$

Here $\Gamma'_{\alpha\beta}$ is the collisional dephasing linewidth, while $\Gamma_{\alpha\alpha}$ and $\Gamma_{\beta\beta}$ are the inverse population lifetimes of the two levels, due to T_1 processes.

In stating that $\Gamma_{\alpha\beta}$ is the observed linewidth, we make the further assumption that the molecule's interaction with the light field is smaller than collisional dephasing linewidths. The molecule-field interaction energy is

$$W = \mu\cdot E \tag{2}$$

where μ is a transition dipole moment and E is the laser electric field. At low fields when $W<<\Gamma_{\alpha\beta}$, the interaction with the light field can be handled by perturbation theory. In this limit, the presence of the field does not change the molecule-liquid eigenstates, but simply introduces transitions among them. This is the normal situation in molecular spectroscopy.

In a three level system involving an excited electronic state (Figure 1), there are typically two types of dephasing process. As electronic wavefunctions are especially sensitive to collisions, the two $S_0\rightarrow S_1$ dephasing rates Γ'_{ac} and Γ'_{bc} are large. However, the ground state vibrational dephasing width Γ'_{ab} is typically much smaller because levels a and b are in the same electronic state S_0.

In the weak field limit, excitation of S_1 produces two kinds of inelastically scattered light in the region of the $S_1 \rightarrow S_0$ (v=1) vibronic transition. As first demonstrated by Y. R. Shen,[2] there is a coherent resonance Raman line of linewidth Γ_{ab} and an incoherent fluorescence component of linewidth Γ_{cb}. The fluorescence line is much broader if electronic dephasing is far faster than vibrational dephasing. This situation is schematically illustrated in the lowest trace of Figure 2.

However, if the field increases such that $\Gamma_{ab} \approx W$, it is no longer possible to use perturbation theory. The dephasing and electromagnetic interactions are then treated on an equal footing by diagonalizing the steady state density matrix. Numerical solution of the model developed by Dick and Hochstrasser[5,6] shows that the resonance Raman and fluorescence appear to "mix" as a function of field intensity. The narrow Raman line first broadens and then splits as a function of W, as shown in Figure 2, if electronic dephasing is larger than vibrational dephasing. In the opposition limit with vibrational dephasing largest, the Raman line first shifts with field intensity.

If a resonance Raman line has a width of ≈ 15 cm^{-1}, then $W \approx 15$ cm^{-1} for a field intensity of $\approx 10^8$ watts/cm^2, assuming resonance with a vibronic transition of oscillator strength near one. This situation has been achieved recently in the ground state spectra of trans-β-carotene in fluid isopentane under intense picosecond excitation.[9] Figure 3 shows the 1525 cm^{-1} ν_1 transition as a function of intensity at

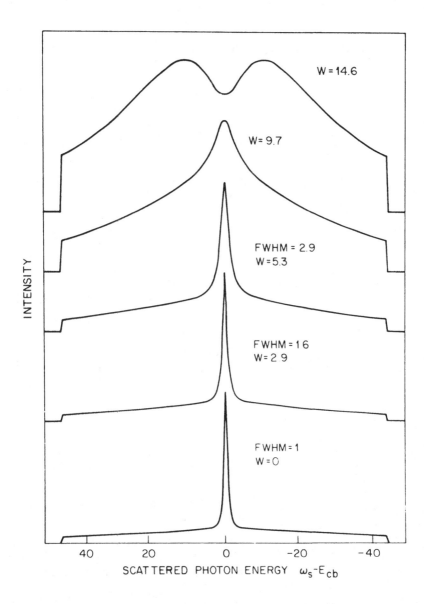

Fig. 2   Calculated inelastic scattering spectra as a function of Rabi energy W, with parameters given in reference 9. All energy scales normalized to the low flux Raman linewidth Γ_{ab}.

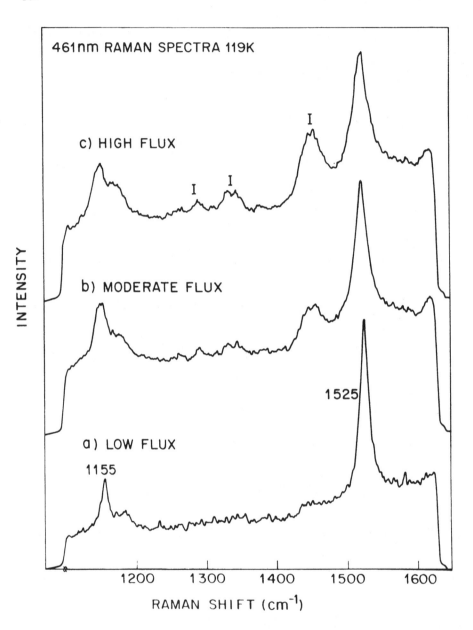

461nm RAMAN SPECTRA 119K

c) HIGH FLUX

I

I   I

b) MODERATE FLUX

1525

a) LOW FLUX

1155

INTENSITY

1200   1300   1400   1500   1600

RAMAN SHIFT (cm^{-1})

Fig. 3   461 nm Raman spectra of β-carotene in isopentane at 119 K as a function

of flux (from reference 9). "I" refers to isopentane lines.

119K. At low field, the lineshape is Lorentzian with a FWHM $\simeq 11 \text{ cm}^{-1}$. As the peak power is raised through the 10^8-10^9 watts/cm^2 region, the line broadens as shown. There is also a saturation observed due to population accumulation in a $\simeq 10$ ps lived, forbidden singlet state, lying below the singlet state in resonance with the 461 nm laser line.

In order to prove that the observed effect is due to high field, and not due to unrelaxed ground state vibrational energy, a simultaneous two pulse - two color ps experiment was carried out. An intense 447 nm pulse partially saturated the transition. The 447 nm Raman spectrum showed broadened lines. In contrast, a 355 nm Raman spectrum generated by a weak simultaneous 355 nm pulse showed no broadening, although the same population saturation (induced by the 447 nm pulse) was observed. As both pulses were generating Raman spectra from the same molecules at the same time, it was concluded that the high field of the 447 nm pulse was responsible for the linewidth.

3. Photochemical Reactions on Semiconductor Crystallite Surfaces

3.1. Crystallite Electronic States and Redox Potentials

The chemical bonding in most semiconductors is quite strong, and the electronic wavefunctions in the valence and conduction bands are delocalized in three dimensions over a wide spatial region. As a consequence, a semiconductor crystallite must grow to fairly large physical size before the electronic bands reach their bulk limiting forms.[10] A recent discovery is the observation that small crystallites (~15Å to 60 Å) made by homogeneous precipitation have the unit cell of the bulk material, but are too small to have bulk band structure.[11] Discrete excited states are instead observed; the crystallite lowest excited state (HOMO to LUMO transition) is shifted to higher energy than the bulk band gap. This size dependent band structure development is schematically illustrated in Figure 4. It is possible to make quantitative models without adjustable parameters for the size dependence of the crystallite MOs.[12-14] The important parameters are the electron and hole effective masses, determined by the curvature of the bulk conduction and valence bands as a function of wavevector.

The figure also shows localized surface states, in addition to internal MOs having nodes on the crystallite surface. The internal MOs evolve into the bulk bands as a crystallite grows. Optical transitions involving the MOs are intense, and dominate the electronic absorption spectra. Evidence for localized surface states comes from luminescence spectra.[14] An optically excited electron in the LUMO can quickly trap in a surface state, and then recombine by tunneling back to a hole in a different surface state. This recombination can be either radiative or nonradiative, and in both cases is strongly coupled to the lattice phonons.

An excited electron in the LUMO, or in a surface state, can alternatively participate in surface redox reactions with adsorbed molecules. While in the bulk crystal the redox potentials of holes and electrons are given by the energies of the valence and conduction band edges, in the crystallite the LUMO electron redox potential will shift negatively from the bulk value by an amount[11]

SPATIAL ELECTRONIC STATE CORRELATION DIAGRAM

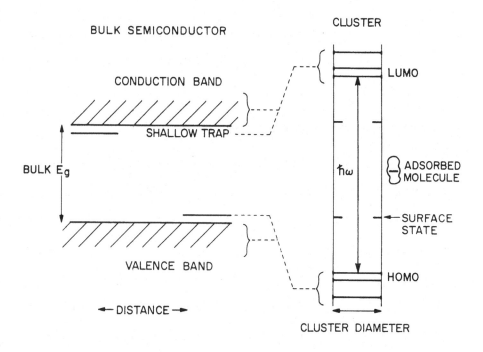

Fig. 4   Schematic illustration of semiconductor cluster electronic states (from

reference 10).

$$\Delta V \simeq \frac{h^2}{8m_e eR^2} \tag{3}$$

Here m_e is the electron effective mass and R is the crystallite radius. The redox potentials of electrons in surface states far below the LUMO level should be less sensitive to size.

3.2. Surface Photoelectrochemistry of Adsorbed Molecules

Surface photoelectrochemistry on TiO_2 and CdS colloidal crystallites has been extensively studied over a period of about 10 years, principally through analysis of final products and via flash photolysis of reacting systems. In applying the time-resolved Raman scattering method to this problem, we attempt to learn how strongly the molecular species must be coupled to the surface in order for electron transfer across the interface to occur on excited state (subnanosecond) lifetimes. If additionally there is a detectable change in the spectra of nasient redox species, then the time evolution of this spectrum yields information on sequential desorption kinetics.

A series of such time-resolved ns and ps resonance Raman studies have been carried out.[15-19] In general the Raman spectrum of the initial charge transfer species shows remarkably little change from the homogeneous liquid spectrum. One of the first reactions reported was oxidation of SCN^- on aqueous colloidal TiO_2 at pH1:[16]

$$TiO_2 \xrightarrow[355nm]{h\nu} TiO_2(e^- + h^+) \xrightarrow{SCN^-} SCN + TiC_2(e^-) \xrightarrow{SCN^-} (SCN)_2^- + TiO_2(e^-)$$

The S=S stretch and its overtone are resonantly enhanced at 532 nm in $(SCN)_2^-$. In order to provide an aqueous reference spectrum for this reaction, $(SCN)_2^-$ made by homogeneous oxidation with electronically excited, triplet ρ-benzoquinone was also studied. Figure 5 shows the low frequency aqueous $(SCN)_2^-$ spectrum from the TiO_2 colloid reaction and from the homogeneous reaction. These two spectra are essentially indistinguishable with respect to peak shape and position. Under the conditions of the experiment, there is a high local concentration of SCN^- on the TiO_2 surface; $(SCN)_2^-$ forms quickly and grows in fully solvated on the ns laser pulse width time scale. A number of surface oxidation and reduction experiments have given similar results; a recent ps experiment on methyl viologen reduction on colloid CdS showed perturbed spectra of the earliest times.[19]

Eosin-y on TiO_2 [18,20,21] provides an example where a photoexcited adsorbed molecule transfers an electron into the crystallite conduction band (i.e., LUMO):

$$E_o \xrightarrow[532nm]{h\nu} E^*(S_1) \xrightarrow[TiO_2]{} E^+ + TiO_2(e^-) \qquad (5)$$

Here E_o is ground electronic state eosin-y, and E^+ is semioxidized eosin. While E_o is soluble in water at neutral pH, at acid pH it protonates, adsorbs on colloidal TiO_2, and undergoes a redshift in its electronic spectrum. The fluorescence quantum yield decreases dramatically upon adsorption; and Grätzel and coworkers report a transient species assigned as E^+ in flash photolysis experiments on the adsorbed species.[20,21]

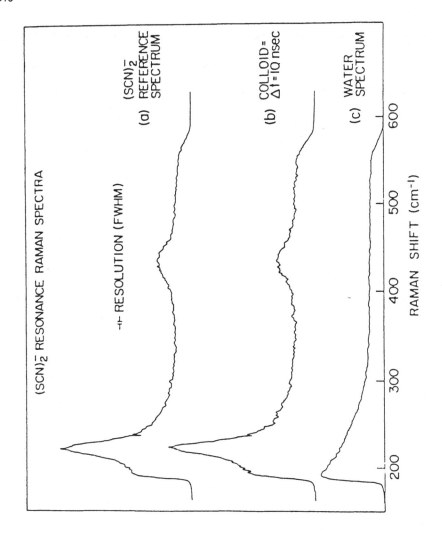

Fig. 5   532 nm resonance Raman spectra of $(SCN)_2^-$ (from reference 16). (a) aqueous reference spectrum, (b) transient spectrum 20 ns following 355 nm excitation of pH1 TiO_2 colloid containing $4 \times 10^{-2} M$ KSCN, (c) underlying H_2O Raman spectrum.

Figure 6 shows, for comparison purposes, E^+ 448 nm resonance Raman spectra in mixed aqueous: alcohol solvents. These spectra show that the intensity ratio of the two modes at 1590 cm^{-1} and 1615 cm^{-1} is a strong function of solvent composition. This sensitivity implies that the degree of resonance between the E^+ excited state and the laser frequency is shifting with solvent. This observation directly demonstrates resonance Raman's sensitivity to local solvation, under the correct resonance conditions.

In colloidal TiO$_2$, ps optical absorption measurements show that the transient species assigned as E^+ appears on a subnanosecond timescale.[21] The calculated quantum yield for eosin singlet decay by charge injection is 38%. In the pump-probe resonance Raman experiments, singlet Eosin is created by 532 nm excitation, followed by a 448 nm Raman probe pulse. In Figure 7 weak yet unambiguous time resolved resonance Raman signals from adsorbed E^+ are observed.[18] The spectrum is a superposition of both E^+ and E_o spectra, as both species show resonance at 448 nm. The spectra confirm the E^+ assignment in the transient optical spectra. The ratio of the 1590 cm^{-1} and 1615 cm^{-1} E^+ transitions is different than observed in water at neutral pH; it suggests, in a general sense, a protonated E^+ in a less polar environment on the surface is being observed. Using ~10 ns laser pulses, the E^+ Raman signal appears without detectable rise time, and decays on a microsecond timescale.

In the Gerischer model for electron transfer across the semiconductor: electrolyte interface,[22] the redox species in solution is assumed to be fully solvated without "specific adsorption." The solvation shell Franck-Condox factors are

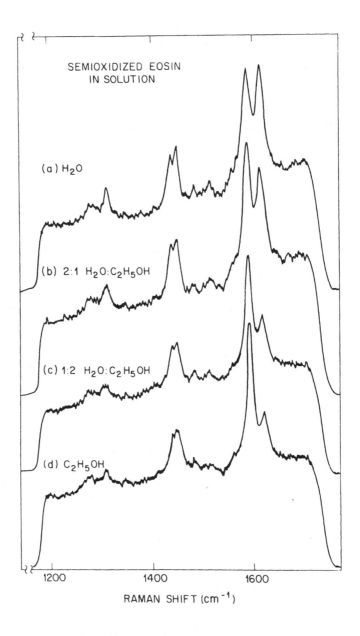

SEMIOXIDIZED EOSIN
IN SOLUTION

(a) H_2O

(b) 2:1 $H_2O:C_2H_5OH$

(c) 1:2 $H_2O:C_2H_5OH$

(d) C_2H_5OH

1200 1400 1600

RAMAN SHIFT (cm^{-1})

Fig. 6 448 nm E^+ resonance Raman spectra as a function of solvent, under conditions described in reference 18.

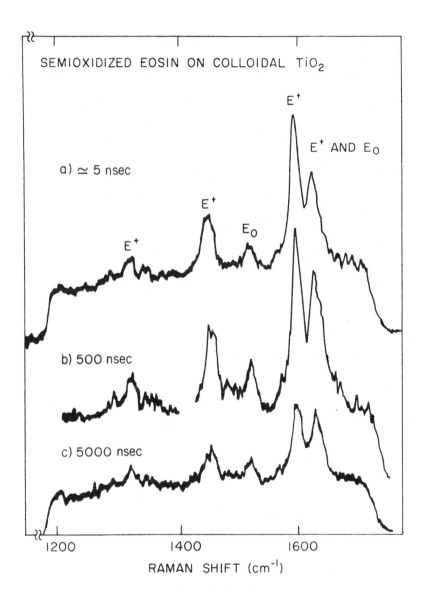

Fig. 7   Time resolved 448 nm E^+ spectra as a function of delay time following

532 nm excitation of adsorbed E_o on colloidal TiO_2 (pH3). (from

reference 18)

significantly involved in the expression for the rate constant. The resonant Raman experiments support this assumption, in that there are no strong deviations from fully solvated spectra, although small changes have been observed in two cases mentioned.

The model considers electron tunneling from plane wave type states in the semiconductor. In the case of methyl viologen reduction on photoexcited colloidal CdS,[19] limited evidence suggests that transfer occurs from an electron trapped in a surface state. In this case, the Franck-Condox factors of atoms in the semiconductor lattice[14] will strongly affect the temperature dependent rate of electron transfer. This aspect should be theoretically incorporated into the electron transfer kinetics.

ACKNOWLEDGEMENT

I am grateful to my collaborators, P. Carroll, R. Rossetti, and S. Beck, who all substantially contributed to the Raman experiments described here.

REFERENCES

1. P. Pagsburg, R. Wilbrant, K. B. Hansen, and K. V. Weisbarg, Chem. Phys. Lett. *39* (1976) 538.

2. Y. R. Shen, Phys. Rev. *B9* (1974) 622.

3. C. Cohen-Tannoudji and S. Reynaud, J. Phys. *B10* (1977) 345.

4. S. Mukamel, D. Grimbert, and Y. Rabin, Phys. Rev. *A26* (1982) 341.

5. B. Dick and R. Hochstrasser, Chem. Phys. *75*, (1983) 133.

6. B. Dick and R. Hochstrasser, J. Chem. Phys. *81* (1984) 2897.

7. Z. Deng and S. Mukamel, J. Chem. Phys. *85* (1986) 462.

8. J. Sue, Y. J. Yan, and S. Mukamel, J. Chem. Phys. *85* (1986) 462.

9. P. J. Carroll and L. E. Brus, "Saturation and Nonlinear Electromagnetic Field Effects in the Picosecond Resonance Raman Spectra of β-Carotene," J. Chem. Phys. *86* (1987).

10. Louis Brus, J. Phys. Chem. *90* (1986) 2555.

11. R. Rossetti, S. Nakahara and L. E. Brus, J. Chem. Phys. *79* (1983) 1086.

12. L. E. Brus, J. Chem. Phys. *80* (1984) 4403.

13. N. Chestnoy, R. Hull, and L. E. Brus, J. Chem. Phys. *85* (1986) 2237.

14. N. Chestnoy, T. D. Harris, R. Hull and L. E. Brus, J. Phys. Chem. *90* (1986) 3393.

15. K. Metcalfe and R. E. Hester, J. Chem. Soc. Chem. Comm. (1983) 133.

16. R. Rossetti, S. M. Beck, and L. E. Brus, J. Am. Chem. Soc. *104* (1982) 7322.

17. R. Rossetti, S. M. Beck, and L. E. Brus, J. Am. Chem. Soc. *106* (1984) 980.

18. R. Rossetti and L. E. Brus, J. Am. Chem. Soc. *106* (1984) 4336.

19. R. Rossetti and L. E. Brus, J. Phys. Chem. *90* (1986) 558-560.

20. J. Moser and M. Grätzel, J. Am. Chem. Soc. *106* (1984) 6557.

21. J. Moser, M. Grätzel, D. K. Sharma and N. Serpone, Helv. Chim. Acta *68* (1985) 1686.

22. S. R. Morrison, *Electrochemistry at Semiconductor and Oxidized Metal Electrodes* (Plenum, New York, 1980), chpt. 3.